Functional Textiles and Clothing 2020

Abhijit Majumdar · Deepti Gupta · Sanjay Gupta
Editors

Functional Textiles and Clothing 2020

Springer

Editors
Abhijit Majumdar
Department of Textile and Fibre Engineering
Indian Institute of Technology Delhi
New Delhi, Delhi, India

Deepti Gupta
Department of Textile and Fibre Engineering
Indian Institute of Technology Delhi
New Delhi, Delhi, India

Sanjay Gupta
World University of Design
Sonepat, Haryana, India

ISBN 978-981-15-9375-8 ISBN 978-981-15-9376-5 (eBook)
https://doi.org/10.1007/978-981-15-9376-5

This Springer imprint is published by the registered company Springer Nature Singapore Pte Ltd.
The registered company address is: 152 Beach Road, #21-01/04 Gateway East, Singapore 189721, Singapore

Contents

Functional Printing and Finishing

Sustainable Production and Supply Chain

About the Editors

Abhijit Majumdar is a Chair Professor in the Department of Textile and Fibre Engineering at Indian Institute of Technology, Delhi. He has more than 23 years of experience in industry and academia. He has published 110 research papers in refereed international journals. His research interests include protective textiles, soft body armour materials, multi-criteria decision making, soft computing, and sustainable supply chain management. He has authored 2 books and edited 3 books, and guided 11 doctoral students. He is a recipient of Outstanding Young Faculty Fellowship (2009-2014) of IIT Delhi, Teaching Excellence Award (2015) of IIT Delhi, Gandhian Young Technological Innovation Award (2017) and FITT award for the Best Industry Relevant Master's Project Supervision, IIT Delhi (2019).

Deepti Gupta is Professor in the Department of Textile and Fibre Engineering atIndian Institute of Technology Delhi. She has more than 30 years of teaching and research experience. Her research interests include ecofriendly finishing of textiles, Functional clothing design, garment sizing and fit. She is a member of various governmental, professional and industrial committees and has edited 4 books, and authored 1 book, 5 chapters and 90 research articles in refereed journals.

Sanjay Gupta is Professor and Vice Chancellor at the World University of Design, Sonepat, India, and was previously the Dean of National Institute of Fashion Technology. He obtained his Ph.D. in textile technology from Indian Institute of Technology Delhi. He was a UNDP fellow to Fashion Institute of Technology (FIT), New York and a Visiting Professor at École Nationale Supérieure des Arts et Industries Textiles (ENSAIT), France. His research interests include Functional clothing design and Development of Textile Products. He has over 100 publications and over 30 presentations in national/ international conferences and seminars.

Testing, Characterisation and Instrumentation

Objective Assessment of the Cooling Function of Textile Products Based on the Combination of WATson and Human Perception

E. Classen

1 Introduction

Comfort is one important issue for clothing. Comfort not only affects the well-being of the wearer, but also his performance and efficiency. Comfort is a complex, highly subjective quality, often defined as the absence of discomfort. The four important aspects of comfort in clothing are thermophysiological comfort, skin sensorial comfort, ergonomic comfort, and psychological comfort [1]. Wear comfort is a complex phenomenon that cannot be properly judged by the customer through simply trying the garment on in the store, nor can it define by sales representative. However, wear comfort can be measured because it is not entirely an undefined, purely subjective individual sensation. Wear comfort is a quantifiable consequence of the body-climate-clothing interaction [2].

Today, the physiological function of textiles and whole garment systems can be measured by a set of laboratory test methods (e.g., sweating guarded hotplate, thermal manikin, sweating thermal manikin). Laboratory test methods are fast and can easily determine different aspects of comfort with high reproducibility. However, the results of laboratory test methods must be correlated with human reception in wearer trials because laboratory tests do not directly measure the comfort.

Clothing with additional functions is more and more important in the field of sports but also in protective clothing. The cooling textiles should support the efficiency of athletes and workers. The cooling effect should improve the comfort and the well-being. During high activity and/or in warm environments the body core temperature can increase and human starts sweating to prevent an overheating of the body. The evaporation of liquid sweat is the most effective process to cool the body. The cooling textile should support the body to keep the body temperature constant. The cooling

E. Classen (✉)
Hohenstein Institute for Textile Innovation gGmbH, Schlosssteige 1, 74357 Boennigheim, Germany
e-mail: e.classen@hohenstein.de

A. Majumdar et al. (eds.), *Functional Textiles and Clothing 2020*,
https://doi.org/10.1007/978-981-15-9376-5_1

effect of textile material is not limited to the use of clothing textiles; the cooling effect is also interesting in the field of bedding, seats and technical textiles.

The cooling of a textile cannot be determined with the conventional test methods of the clothing physiology. To determine the cooling power of fabrics, the new heat release tester WATson was developed in Hohenstein. With WATson, the cooling power of cooling materials can be determined and compared. However, the measured cooling power is only a physical value. Without the correlation of these values with data of subject trials, the cooling power does not give any information about the perception of the human body and the achieved cooling effect.

2 Materials and Methods

2.1 Materials

The textile materials were commercially available fabrics for sport wear (see Table 1). The material composition was taken from the product information of the manufacturer. The mass per unit was determined according to DIN EN 12127. The thickness was determined according to DIN EN ISO 53855.

All fabric samples were washed once prior to testing on household washing, 30 °C, line drying with a standard detergent (ECE-2 Standard Detergent 1998) in accordance with ISO 6330 (procedure 4M for polyester (PES), polyamide (PA) and polypropylene and procedure N for cotton (CO)).

For sample preparation, three specimens measuring 25 cm × 25 cm were cut out. The test specimens were conditioned for 12 h prior to measurement under the climatic conditions of the testing protocol.

2.2 WATson

The test device WATson is placed in a climate chamber to ensure constant and defined ambient climate during the test run (ambient temperature (T_a) of 30 ± 0.5 °C and relative humidity of in the climate chamber (rH_a) of 70% ± 10%). The evaporative heat loss test device WATson consists of a heated plate with sweat glands and is technically and electrically designed to mimic human thermoregulation (see Fig. 1). The area of the measuring head is 400 cm^2 (20 cm × 20 cm). The heated plate is set to an average skin temperature (T_s) of 32 ± 0.1 °C. The temperature of the measuring head is held constant at the set temperature by controlled electrical heating. This electrical heating power to maintain this set temperature (i.e., the heat loss) is recorded at 1 datapoint per second (1 Hz) and is stated as "$P_{heating}$" in W.

The four inner sweat glands supply deionized water (i.e., "sweat") via a peristaltic pump with a pumping rate (i.e., sweat rate) of 8 ± 0.1 g/h. Wind is one important

Tab 1 Materials and characteristic parameter

Sample	Composition*	Mass per unit (g/m^2)	Thickness (mm)	Construction
M01	PES (100)	120	0.5	Knitted, mesh
M02	PES (100)	95	2.4	Spacer fabric
M03	PA/PES/El/Lycra (54/32/14)	245	0.85	Knitted, two layers
M04	PA/PES/El/Lycra (58/26/16)	240	0.93	Knitted, two layers
M05	PA/PES/El/Lycra (58/29/13)	250	0.88	Knitted, two layers
M06	PES/Carbon (98/2)	120	0.43	Knitted
M07	PES/PES Coolmax (50/50)	120	0.75	Knitted, two layers
M08	PES/EL/Lycra (67/33)	220	0.76	Knitted
M09	PES (100)	119 120	0.69	RR double knit
M10	PES (100)	119 113	0.67	RR double knit
Fabric 1	PES (100)	**	**	Knitted
Fabric 2	PES (100)	**	**	Knitted
Fabric 3	PES (100)	**	**	Knitted

*Composition of fabric from manufacturer information; PA = polyamide, PES = polyester, EL = elastane, **fabric 1,2 and 3 show the same mass per unit and the same construction

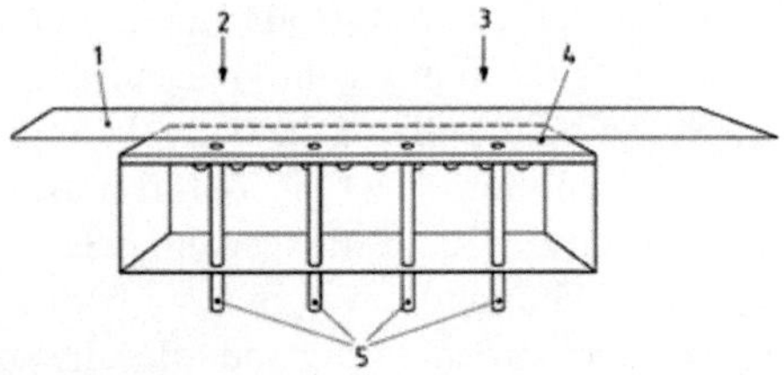

Fig. 1 Schematic illustration of the evaporative heat loss test device WATson

1 Textile Sample
2 Wind, strictly perpendicular and centric
3 IR radiation (simulating solar radiation), not perpendicular
4 Measuring head
5 Sweat glands

parameter of the real wearer situation and for wind simulation, the test was done by a light breeze of 1 ± 0.1 m/s. The solar radiation is the second important parameter of the real wearer situation and was simulated by an IR-lamp with 13.2 ± 0.1 W. The textile sample is placed in direct contact with the heated plate.

As a link to reality, this heating power equates the heat loss of the skin (identical with the heat loss of the fabric) due to the evaporation of sweat and can be described as the ability to lose evaporative heat when wearing this kind of clothing. So, the higher this heating power, the higher is the physiological cooling effect, that is, the cooler the fabric is perceived on the skin.

2.3 Wearer Trials

To determine the influence of cooling textiles on human thermoregulation and temperature perception wear trials with subjects were performed. Test design followed ethical rules and written informed consent was obtained from all participants.

Five male subjects (28.8 ± 3.2 years, 178.8 ± 7.3 cm, BMI 23.6 ± 1.1) performed a standardized activity protocol in a climatic chamber (temperature T_a = 25 °C, relative humidity RH_a = 50% rh). Subjects were running on a treadmill with different speeds: 4 and 6 km/h. They were wearing cooling shirts and additional standard clothing (short pants, socks, shoes). Samples for the cooling shirts were chosen by the results of the heat release tester. Every subject performed three trials with one t-shirt. Samples were pre-acclimatized for at least two hours. Trial duration was 120 min with a activity and rest program (0–20 min: sitting on a chair (rest); 20–60 min walking with a speed of 4 km/h; 61–80 min: sitting on a chair (rest) and 81–120 min walking with a speed of 6 km/h.

T-shirts were made from three different cooling textiles with different cooling behavior (fabric 1, 2 and 3). To produce the T-shirts of the wearer trials enough material had to be available; this was only the case of the fabric 1, 2 and 3. All other investigated materials were ready-made products and could not use in the wearer trials. For a fit of the T-shirts, the size and shape of the test subjects were determined with 3D-Scanning to achieve the body data for the sewing of well-fitted T-shirts. To achieve a high cooling effect the T-shirt must be worn closed to the body of the test subjects. Objective data of the test subjects (e.g., heart rate, core temperature, skin temperature and humidity, weight loss) with sensors and data logger and subjective feedback were recorded during and after the subject trials.

3 Results and Discussion

Figure 2 shows the results of 10 investigated products from the market with the heat release tester. The samples were put on the WATson measuring head in the dry state. Sweating was switched on at t = 10 min and performed until a constant heating power was achieved again (i.e., heat loss in wet state). Then sweating was turned off (t = 70 min) and the test was performed until the samples were dry again (i.e., drying time, the decay of heat loss over time).

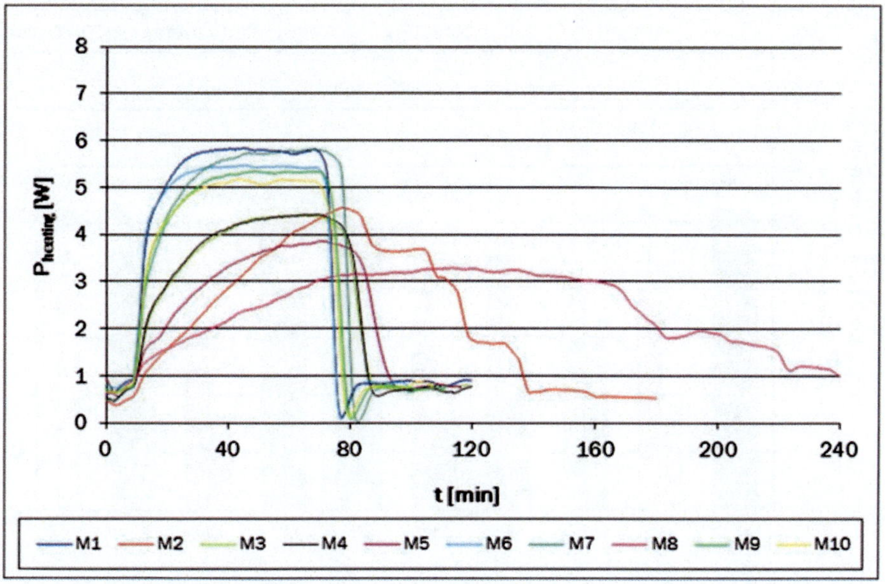

Fig. 2 Heating power of cooling textiles M1–M10, (WATson results)

The ten investigated textiles show different curves. Fabric M1, M6, M7, M9 and M10 show a fast increase in the heating power and reach the maximum cooling power fast. The drying period starts also very fast at the end of sweating and the fabrics show a similar drying behavior. The other fabrics show a low increase of the cooling power in the first 20 min, a lower maximum cooling power and the drying phase is very various. Fabric M2 and M8 show the longest drying times. Important for the cooling is a fast increase of the cooling power after the start of sweating and the reached maximum cooling power. For a better comparison, the heating power at certain periods was investigated and the results show that certain time periods for the cooling behavior are important: the wicking power, the cooling power and the drying behavior. These aspects are shown in Figs. 3 and 4.

Figure 3 shows the so-called wicking power and the cooling power of the different fabrics. After analysis of the data, the wicking power can be expressed as the momentary value of the heating power $P_{heating}$ at t = 20 min.

$$P_{wicking} = P_{heating}(t = 20) \tag{1}$$

The cooling power is the average of $P_{heating}$ between t = 60 min and t = 70 min.

$$P_{cooling} = \frac{\sum_{i=1}^{n} P_{heating,i}}{n} \tag{2}$$

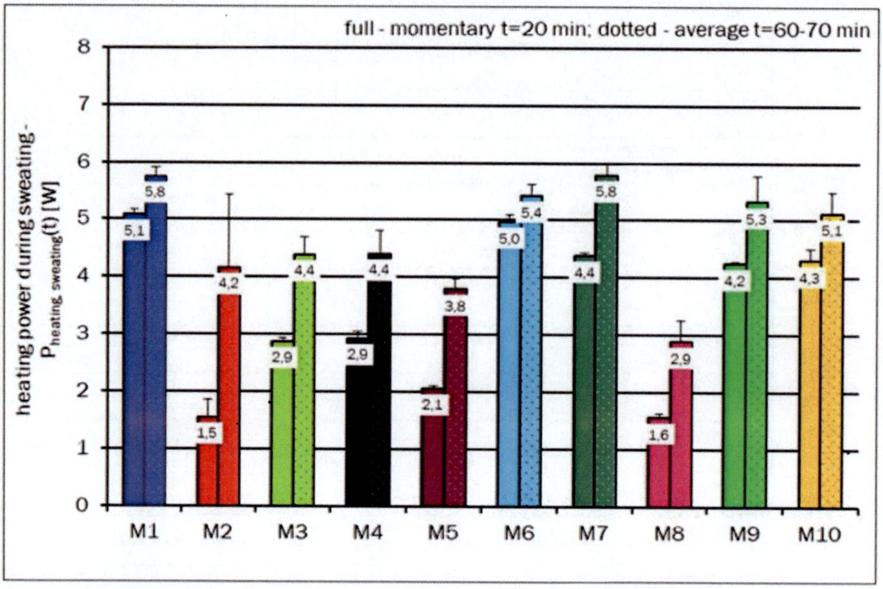

Fig. 3 Heating power of different cooling textiles, at 20 min and average value over 10 min between 60 and 70 min (results WATson)

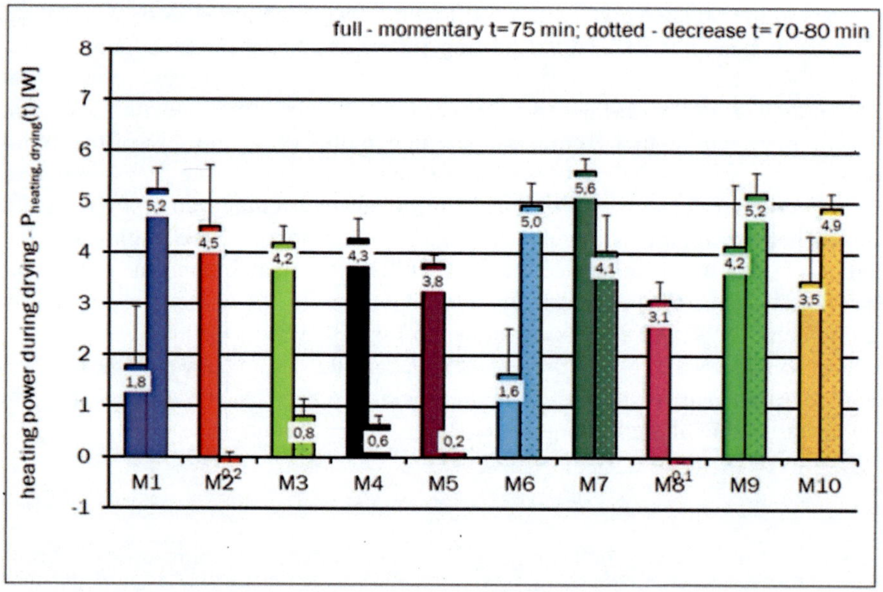

Fig. 4 Heating power of different cooling textiles, at 75 min and average value over 10 min between 70 and 80 min (results WATson)

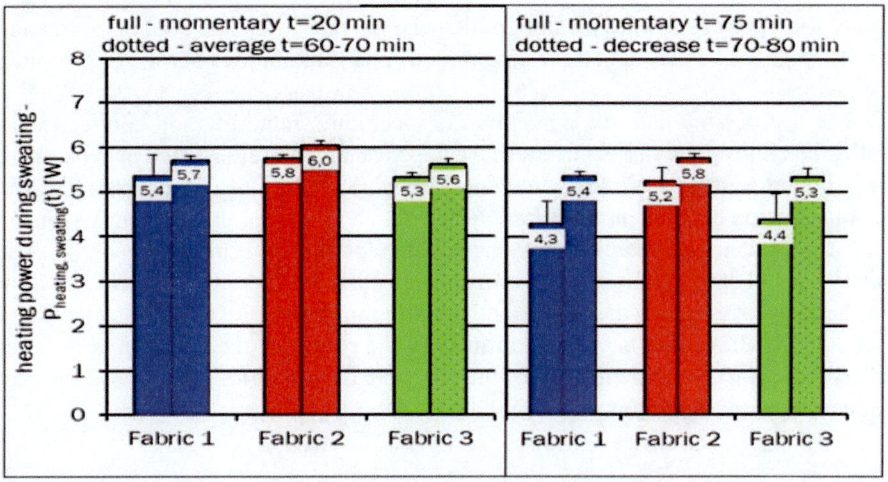

Fig. 5 Heating power of fabric 1, 2 and 3, at 20 min and average value over 10 min between 60 and 70 min (sweating phase) and at 75 min and average value over 10 min between 70 and 80 min (drying phase)

n is the number of measurements, that is, n = 600 for measurements in a time interval of 10 min, if $P_{heating}$ is measured every second.

The highest heating power at the beginning of the sweating (t = 20 min) shows the fabric M1, followed by fabric M6, M7, M9 and M10. The highest cooling power (t = 60–70 min) shows the Fabric M1, followed by fabric M7, M6, M9 and M10. Fabric M2 shows the lowest heating power, in the beginning, followed by fabric M8, M5, and M3 and M4. The highest cooling power show fabric M3 and M4, followed by fabric M2, M5 and M8. Fabric M9 and M10 show the same material composition and construction; however the cooling behavior show differences. The reason for the different cooling behavior could be, for example, the different fiber shape of the PES fibers. Further investigation is necessary to clarify the reason for different cooling behavior.

The analysis of the drying behavior shows that two periods are important for the comparison of the cooling power: the heating power at 75 min and the average heating power between 70 and 80 min. The drying time is the time starting from t = 70 min until t = x min; x is reached when $P_{heating}$ reaches the same value as at t = 10 min. The drying performance/drying time of a tested sample was displayed as a momentary value at the 75th min of the experiment (i.e., 5 min after the end of sweating) and the average over the first 10 min of drying phase-out of the sample (Fig. 4). The lower the value at the 75th minute and the higher the average value between 70 and 80th minute the quicker the fabric dries (quick-drying fabrics). The higher the value at the 75th minute and the lower the average value between 70 and 80th minute the longer the fabric dries (slow-drying fabrics).

Figure 5 shows the different characteristics parameter of the fabric 1, 2 and 3 measured with the heat release tester WATson. These fabrics were used in the subject trials. The fabrics show high values of the wicking power and the cooling power and the average value over a certain time period. The three fabrics show good cooling behavior.

The test persons rate the perception (temperature, humidity and comfort) on a rating box during the subject trials. The subjective temperature is rated by a modified seven-point Bedfort scale [3]: 1 too cold, 2 cold, 3 slightly cold, 4 neutral, 5 warm, 6 hot and 7 too hot. For humidity, a four-level scale is used: 1 dry, 2 slightly moist, 3 moist and 4 wet. The details of this analysis are reported in the project report [4]. The analysis of all data which are received from the wearer trials show that fabric 2 tends to be the best during and in the end the activity program in the temperature and humidity perception. These results were confirmed by the results of the questionnaires after every subject trial and the results of the objective data. Fabric 2 shows also the best values in the investigation with WATson.

4 Conclusion

Differences between cooling textiles could be shown by means of heat release tester WATson and wear trials. WATson test design can show differences in the heating power of different cooling textiles during the test program. The test program simulates the human sweating during activities in warm conditions (temperature 25 °C, relative humidity 50 %) and the time after activities. The first 10 min simulates the non-sweating phase (pre-exercise state) representative for low metabolic rates without sweating. After 10 min, the sweating is starting for 1 h. This represents high metabolic rates with liquid sweat occurring and equates the evaporative heat loss of the human skin covered with the wet test specimen. After 70 min, the sweating is stopped, and this represents the activity by rest with low metabolic rates. This equates the wet heat loss of the human skin covered with the test specimen in wet to dry transition and the drying time of the test specimen on the human skin. The evaporative cooling power of a tested sample in the sweating phase can be displayed as a total average over the sweating phase and as a steady-state average over the last 10 min of the sweating phase. The closer these 2 bars are together the quicker the build-up of the maximum evaporative cooling power is. The further these 2 bars are apart the longer it takes to build up the maximum evaporative cooling power. The drying performance after activity can be displayed as a momentary value at the 75th min of the test and the average over the first 10 min of drying out of the sample. The lower the value at the 75th minute and the higher the average value between 70 and 80th minute the quicker the fabric dries. The higher the value at the 75th minute and the lower the average value between 70 and 80th minute the longer the fabric dries. The wearer trials of three different garments show that the results of objective and subjective parameters of the tested garments are similar and fabric 2 is judged as the best one which is seen in the WATson test. Because the three tested fabrics have similar properties further wearer trials with cooling textiles will be performed in the future to correlate the data of wearer trials.

References

1. Mecheels J (1998) Körper-Klima-Kleidung: Wie funktioniert unsere Kleidung? Schiele & Schiele, Berlin
2. Classen E (2018) Advanced characterisation and testing of textiles. In: Dolez P, Vermeersch O, Izquierdo V (eds) Woodhead publishing, Elsevier Ltd.
3. Harnisch M, Katz B (2010) Improvement of the physiological function of sport wear by comfort zones. Final report of the IGF research project 15720N (German). Hohenstein Institut fuer Textilinnovation, Bönnigheim, Germany
4. Classen E (2019) Development of a clothing physiological model for the evaluation of cooling textiles. Final Report of the IGF research project 18292N (German), Hohenstein Institut fuer Textilinnovation. Boennigheim Germany, September 2019

Validation and Reliability of Sizestream 3D Scanner for Human Body Measurement

Manoj Tiwari and Noopur Anand

1 Introduction

Knowledge of anthropometric body dimensions, size, shape, movement, etc., is of much use in many fields. Traditionally, such data have been obtained through manual techniques of measuring the body through hand-held instruments like anthropometer, measuring tapes, spreading caliper, etc., which is time-consuming, labor-intensive and hence expensive. It further becomes more challenging when a data has to be collected on a larger scale from many participants like in national sizing surveys. Additionally, the data such collected (manually) is error-prone and hence unreliable. This is the reason why there has been a significant increase in the utilization of 3D body scanning technologies worldwide for such purposes. More than 19 countries since 1980s have undertaken their national anthropometric surveys using 3D scanning technology for example the USA, the UK, Germany, France, Japan, Korea, and China, etc. The 3D scanning has applications in a number of fields such as automobiles, medical sciences, aviation, architecture, and of course apparel where it is extensively used for mass customization for an improved fit.

M. Tiwari (✉)
Department of Fashion Technology, National Institute of Fashion Technology, Jodhpur 342037, Rajasthan, India
e-mail: manoj.tiwari@nift.ac.in

N. Anand
Department of Fashion Technology, National Institute of Fashion Technology, New Delhi 110016, India

A. Majumdar et al. (eds.), *Functional Textiles and Clothing 2020*,
https://doi.org/10.1007/978-981-15-9376-5_2

2 Background

Since the collection of anthropometric data automatically through 3D whole-body scanning technology is getting favored over traditional manual measurement method, it becomes vital to establish the accuracy of the new technology-driven system in comparison to the traditional manual system for establishing acceptability and appropriateness of the data generated. It is important to understand how the scanner extracted measurements are concurring with the measurements taken manually for the same dimensions [1]. The validity and reliability assessment of 3D body scanner is one critical area in order to establish the accuracy of the data collected. There have been a number of researches conducted in this aspect of 3D scanning which is further supported by ISO. Further, there have been recommended standard procedures to establish 3D scanner performance in terms of reliability and accuracy, in different ISO documents as well. ISO 20685:2005 guides on 3D scanning methodologies for internationally compatible anthropometric databases [2] while ISO 20685:2015 discusses about the evaluation protocol of surface shape and repeatability of relative landmark positions [3]. This study is aimed at establishing the validity and reliability of 3D whole-body scanners to carry out an anthropometric survey.

3 Methodology

3.1 Procedure and Equipment

Sample. Fifty (50 in no.) subjects (25 males and 25 female subjects) of different body shapes/body types were measured manually as well as using 3D whole-body scanners.

Anthropometric measurements were manually collected through traditional techniques by measurers. In total twelve (12) body dimensions comprising of height, girth (small and big), body depths were shortlisted to be measured as part of the study namely, Stature, Waist back height, Shoulder height, Inside leg-length, Waist height, Across-shoulder width, Chest girth, Waist girth, Hip girth, Neck girth, Thigh girth, and Chest. It was ensured that at least one dimension was included from the ISO 20685:2005 prescribed measurement type [2]. It may be noted that dimensions related to head, hand and foot were not in the scope of this study, hence were not included in the scanner validation exercise.

The manual body measurements were taken manually by the expert having experience of anthropometry. There were two (02) measurers to measure male and female subjects. Each of the measurer measured every subject twice independently following the standard definition and measurement procedure as prescribed in ISO 8559:1989. The manual body measurements (lengthwise) were taken using an anthropometer and the sliding stadiometer of length 210 cm and least count 1.0 mm. While for the girth related measurements, a certified flexible non-stretchable steel tape (Length

200 cm and least count 1.0 mm) was used. A mirror behind the subject was fitted on the wall to ensure the correct landmark positioning and placement of measuring tape. The entire exercise of taking manual body measurements was conducted under the guidance of an observer having more than 20 years of experience in anthropometry and anthropology.

Anthropometric measurements are automatically collected through the use of 3D whole-body scanning technology. The technology used was infrared technology by SS14 3D Body scanning Sizestream. The protocol followed while 3D scanning was as per ISO20685:2005. The scanner was duly calibrated at the start of the exercise on each day as per the procedure prescribed by the manufacturer. Each of the subject was measured thrice using 3D Scanning (1 scanner scanned each of the subject 3 times). To ensure precise 3D body measurement while scanning, each of the subjects were provided specially designed scan suits made of a material with enough stretch to avoid any body compression as well as any kind of slackness or looseness from the body. During 3D scanning process, the subjects were asked to maintain the posture by holding their breath in an exhaled position as prescribed in ISO 20685:2010-11.

This resulted in a total of seven (07) data points for each of the body dimension (03 data points from 3D scanning as each of the subject was scanned thrice, and 04 data points from manual scanning, wherein each of the subject was manually measured twice by each of the two measurers).

3.2 Data Preparation

Step 1. 3D scans (done thrice) for each of the subject.

Step 2. Manual measurements (done twice by each of the two measurers) for each of the subject.

Step 3. Recording and storing 3D scan data and manual measurement data of each subject (for all the prescribed dimensions) in the MS Excel sheet/SPSS file.

3.3 Data Analysis

Step 1. Differences between the 3 measures derived from the scanner and 4 measures provided by the manual measures were computed. There are 12 combinations of the same for each measure and for a particular subject. The mean difference was calculated by averaging all the differences between the scanner measurements and the average measurements taken manually [1]. The differences were checked using the test methods given in Clause 5 of ISO 20685:2005 [2, 4].

Standard Deviation (SD) of the difference between scanner measurements and the manual measurements to check the comparability of the two measurement techniques was calculated [5].

Step 2. Error limits were calculated using the mean of the differences (between scanner measurement and manual measurement) for all the subjects and reported with its associated standard deviation, sample size and 95% confidence Interval. If the 95% confidence interval for the mean of scan-minus-manual measure differences is within the plus or minus interval defined by the ISO 20685:2005 values, then the 3D scanning system can be considered as acceptable as per the International standards [6].

$$\text{95\% Standard Error} = 1.96 * \text{std. Deviation of difference/SQRT(N)}$$
$$\text{Upper Limit} = \text{Mean difference}(+)\text{95\% Standard Error}$$
$$\text{Lower Limit} = \text{Mean difference}(-)\text{95\% Standard Error}$$

Please refer Table 1 for the scanner validation analysis on the actual extracted measurements from the scanner. Table 1 illustrates the summary statistics for scan measurements-manual measurements for different body dimensions measured during the scanner validation exercise. The upper limit and lower limits of the differences as derived by following the process mentioned in step 2 were compared to the ISO 20685:2005 standard values for the permissible error. The result of the comparison of error values obtained from scanner measurements to the ISO 20685:2005 permissible error is shown as "Outcome" with P for Pass or F for Fail. It can be observed that the upper and lower limit of all the differences were observed beyond the permissible limit except WBL (Waist back length) and WH (Waist height), where the lower limit and the upper limit was observed within the permissible range for WBL (Waist back length) and WH (Waist height), respectively.

To accept the scanner extracted measurements, both the upper value and lower value should be within the ISO 20685:2005 prescribed limit. For example, for Stature the upper limit and lower limit are 1.42 and 1.17 cm, respectively, while the ISO 20685:2005 permissible error value is ± 0.4 cm. Here both the values of the upper limit and lower limit are beyond the ISO 20685:2005 permissible error value, hence all the scanner extracted measurements were considered as "FAIL".

Further, the same procedure was followed while checking the accuracy as per ISO 8559:1989 standard by comparing the difference between scanner measurements and manual measurement to the ISO 8559:1989 values (an accuracy of ± 1% or ± 5.0 m, whichever is the smaller) [7]. As indicated in Table 1, While comparing the mean difference of scanner measurement – manual measurements to the ISO 8559:1989 prescribed error limits, it was observed that all the differences were observed beyond the permissible limit. However for the WH (Waist height), the upper limit was observed in the ISO Range (Error of −0.43 against standard acceptable error value of −0.497) and a lower limit was observed out of the ISO Range (Error of −1.02 against standard acceptable error value of −0.497), hence all the scanner extracted measurements were considered as "FAIL".

Table 1 Scanner validation (Actual Scanner Extracted Measurements)

3D Scanner Validation (on Actual Extracted Scanner Measurements)												
Measurement	Stature	Waist back length	Shoulder height	Inside leg-length	Waist height	Across shoulder	Chest girth	Waist girth	Hip girth	Neck girth	Thigh girth	Chest depth
Mean Diff (SCAM-MAM) (cm)	1.294	0.812	−1.382	−2.088	−0.723	3.947	1.927	3.292	2.017	2.648	1.564	0.999
SD. Of Diff (cm)	1.518	3.914	1.889	1.541	0.6293	3.256	2.024	5.369	2.085	3.978	2.373	1.386
Error @95% (±)	0.123	0.316	0.153	0.125	0.293	0.263	0.164	0.434	0.169	0.322	0.192	0.112
Upper limit	1.42	1.13	– 1.23	– 1.96	– 0.43	4.21	2.09	3.73	2.19	2.97	1.76	1.11
Lower limit	1.17	0.50	– 1.53	– 2.21	– 1.02	3.68	1.76	2.86	1.85	2.33	1.37	0.89
ISO 20685:2005 Accept. Limit (cm)	0.400	0.500	0.400	0.400	0.400	0.400	0.900	0.900	0.900	0.400	0.400	0.500 ara>
Outcome*	F	F	F	F	F	F	F	F	F	F	F	F
Accept. Error Based on ISO 8559:1989, % values (cm)	0.500	0.371	0.500	0.500	0.500	0.314	0.500	0.500	0.500	0.367	0.500	0.238
Outcome	F	F	F	F	F	F	F	F	F	F	F	F

* P: Pass, F: Fail, SCAM: Scanner Measurement, MAM: Manual Measurement

Step 3. The results of the data analysis conducted in step 2 showed that the error of scanner extracted measurements were observed beyond the prescribed acceptable limits of ISO 20685:2005 and ISO 8559:1989. However, it was observed that the error in the scanner extracted measurements was unidirectional.

Intra-class correlation coefficients (ICC) was conducted to check for the consistency and repeatability between different scan measurements and it represented the variance attributable to error [8, 9]. ICC reflects both degree of correlation and agreement between measurements. Values less than 0.5 are indicative of poor reliability, values between 0.5 and 0.75 indicate moderate reliability, values between 0.75 and 0.9 indicate good reliability, and values greater than 0.90 indicate excellent reliability [10]. The ICC between the different scan measurements (for a given dimension) were calculated using IBM SPSS Statistics 23.0. Please refer Table 2 for ICC values of scanner measurements between one to other.

As indicated in Table 2, the higher values of ICC (> = 0.90) for each of the dimension except for Neck girth. The reason for low ICC values for neck girth may be due to the disturbance in scanning caused by hair, especially with female subjects. This might have caused an error while the scanner identifies landmark to measure the neck girth. The ICC values for male subjects were observed good (0.912, 0.947, and 0.903) and confirmed a good level of consistency and reliability.

Bias-shift. As the scanner measurements were observed consistent and repeatable, a correction factor in the form of bias-shift was incorporated into the scanner extracted measurements. The concept of bias-shift is like a tare function were going forward, the average difference between scanner measurement and the manual measurement (as a standard correction value associated with specific body dimension) was adjusted to the scanner extracted measurement. It is suggested that the two measurements would become comparable by reducing the bias (mean) and/or random fluctuation (SD) of errors, however, reducing mean error (bias) is more effective than reducing SD to make the two measurements comparable [11]. A bias is a noticeable systematic error in the scanner measurements, which can be used as an offset to correct the scanner measurements in order to achieve improved concordance [12, 13].

Impact of applying bias-shift on the accuracy level achieved against ISO 20685:2005 prescribed error levels was checked by repeating the validation analysis (as explained in step 2) after applying bias-shift as a correction factor. It was observed that after incorporating the bias-shift, all the differences between the scanner and manual measurements were observed within the ISO 20685:2005 prescribed acceptable limits (refer Table 3 for Summary results after incorporating bias-shift), hence the scanner measurements could be considered comparable against the manual measurements, which were used as prescribed gold standards for checking the accuracy. Hence, the scanner extracted measurements were considered as "PASS".

Following the same procedure, the average difference between scanner measurements and the manual measurements for each dimension was also compared to the ISO 8559:1989 prescribed limits. As indicated in Table 3, it was observed that the error between scanner measurements and the manual measurements for all

Table 2 Intra-class Correlation (ICC)- Scanner Measurements

Dimension	Stature	Waist back length	Shoulder height	Inside leg length	Waist height	Across shoulder	Chest girth	Waist girth	Hip girth	Neck girth	Thigh girth	Chest depth
ICC (SCAN1–SCAN2)	0.995	0.905	0.997	0.988	0.953	0.989	0.999	0.978	0.998	0.384	0.985	0.998
ICC (SCAN1–SCAN3)	0.993	0.922	0.996	0.976	0.939	0.983	0.998	0.956	0.997	0.419	0.978	0.997
ICC (SCAN2–SCAN3)	0.999	0.937	0.997	0.989	0.948	0.988	0.998	0.968	0.998	0.865	0.988	0.998

Table 3 Scanner validation (After incorporating Bias-Shift)

3D scanner validation (After incorporating Bias-Shift)												
Measurement (BIAS-SHIFT) Combined—MALE–FEMALE	Stature	Waist back length	Shoulder height	Inside leg-length	Waist height	Across shoulder	Chest girth	Waist girth	Hip girth	Neck girth	Thigh girth	Chest depth
Bias-shift applied (cm)	– 1.2935	– 0.8123	+ 1.3821	– 2.0881	+ 0.7228	– 3.9474	– 1.9270	– 3.2917	– 2.0171	– 2.6476	– 1.5642	– 0.9987
Mean Diff (SCAM-MAM) (cm)	0.000	0.000	0.000	0.000	0.000	0.000	0.000	0.000	0.000	0.000	0.000	0.000
SD. Of Diff (cm)	1.518	3.914	1.889	1.541	3.629	3.256	2.024	5.369	2.085	3.978	2.373	1.386
Error @95% (±)	0.123	0.316	0.153	0.125	0.293	0.263	0.164	0.434	0.169	0.322	0.192	0.112
Upper limit	0.10	0.32	0.15	0.10	0.29	0.26	0.16	0.43	0.17	0.32	0.19	0.11
Lower limit	– 0.10	– 0.32	– 0.15	– 0.10	– 0.29	– 0.26	– 0.16	– 0.43	– 0.17	– 0.32	– 0.19	– 0.11
ISO 20685:2005 Acceptable Limit (cm)	0.400	0.500	0.400	0.400	0.400	0.400	0.900	0.900	0.900	0.400	0.400	0.500
Outcome*	P	P	P	P	P	P	P	P	P	P	P	P
Acceptable Error Based on ISO 8559:1989, % values (cm)	0.500	0.371	0.500	0.500	0.500	0.314	0.500	0.500	0.500	0.367	0.500	0.238
Outcome	P	P	P	P	P	P	P	P	P	P	P	P

* P: Pass, F: Fail, SCAM: Scanner Measurement, MAM: Manual Measurement

the dimensions was within the ISO 8559:1989 limits. Hence, the scanner extracted measurements were considered as "PASS".

In view of the observed difference between measurement reported by the scanner and one took manually, the same set of subjects were scanned by two other SS14 scanners, under the same scanning conditions, to examine scanner consistency and eliminate the possibility of scanner misreporting. It was seen that similar results (i.e. deviations) were reported from all the three scanners. Intraclass correlation ICC was calculated between the measurements extracted from all the three scanners and the result of ICC ranged between 0.997 to 1.00 for all the anthropometric measurements. This concludes that scanners had high levels of consistency with a systematic error. This pattern has been observed with other scanner technologies as well.

Han et al. [5] conducted a comparative analysis of 3D body scan measurements and manual measurements of Size Korea adult females using 3D whole-body laser Scanning technology of Cyberware (WB4) and observed that scanner measurements were generally larger than manual measurements with that the mean differences of circumferences being larger than those of lengths and heights [5].

Mckinnon and Istook [1] undertook a similar comparative study of 10 anthropometric measurements of dress forms extracted from two 3D whole-body white light scanning technology i.e. the Image Twin (2T4) and the 3T6, with the measurements obtained manually and reported that the scanned measurements from both the scanning systems exceeded the manual measurements except crotch circumference which was found consistently smaller [1].

Gordon et al. [14] in Anthropometric Survey of U.S. Army Personnel- Methods and Summary Statistics report (ANSUR II conducted using 3D whole-body laser Scanning technology Cyberware WBX 3D) stated that on comparison of measurement extracted through traditional anthropometry methods with the one extracted using 3D Scanning technology, scan-generated measurements were observed to be significantly larger [14].

Koepke et al. [9] while comparing 3D laser-based photonic scans (using 3D whole-body laser Scanning technology of Human Solution—Vitus Smart XXL) and manual anthropometric measurements for Swiss men reported the scanned derived values of height were found systematically shorter by 2.1 cm, while the waist, hip and bust circumferences were observed larger in the scans by 1.17–4.37 cm [9].

Šimenko and Čuk [15] in testing Reliability and Validity of 3D whole-body white light technology of TC2 NX-16 confirmed that 3D body scan values were larger but systematic with a high level of consistency [15].

There have been many such similar studies by Kouchi [6], Kuehnapfel et al. [12], Wang et al. [16], Wells et al. [17], Zancanaro et al. [18], Bougourd et al. [19], Paquette et al. [20], Robinettea and Daanend [21], and Weinberg et al. [22] where comparison of 3D scan extracted measurements to manual measurements using various 3D whole-body scanning technologies have been done and deviations recorded from those of manual measurements. These researches have also confirmed that the scanned derived measurements show a high level of consistency in comparison to manual measurements [6, 12, 16–22].

This established the fact that the scanner though being very consistent with scanned measurement reporting may have a systematic error that can be statistically corrected. One such method for correcting the difference between scanner reported measurement and manual measurement is Bias shift method as proposed in this paper.

4 Results and Discussion

It was observed that the Sizestream SS14 scanners used in the validation exercise were highly consistent in taking human body measurements, however, it indicated a systematic error throughout the process which lead to failure of all the measurements in terms of accuracy levels (as per ISO 8559:1989 and ISO 20685:2005) achieved against manual measurements used as the gold standards. Subsequently, a correction factor in the form of bias-shift was determined, and the same was incorporated into the scanner extracted measurements and scanner measurement accuracy was achieved.

5 Conclusion and Future Scope

This research paper explains the detailed methodology followed to establish the accuracy and reliability of the Sizestream 3D full-body scanning system. It also describes the statistical procedure to confirm the measurement consistency of the scanners. The scanner high level of consistency (with-in and between) confirmed the behavior of systematic error across all the SS 14 body scanners used in the study. As the scanners were highly consistent, the same correction factor (in form of Bias-shift) may be applied to all other SS 14 body scanners. The application of bias-shift (as a correction factor) confirmed the scanner measurements within the permissible error limits of ISO 20685:2005 and ISO 8559:1989. The 3D scanner validation methodology discussed in this paper may be applied successfully in future researches related to 3D scanning recommended to conduct sizing surveys.

Acknowledgements The authors are thankful to Prof. Vandana Narang, Prof. Suhail Anwar, Prof. Monika Gupta, and Prof. S.S. Ray of National Institute of Fashion Technology, India for their technical inputs to carry out this research. We are also highly thankful to the participants for their volunteer contribution as subjects for 3D scanning in this study.

References

1. Mckinnon L, Istook C (2001) Comparative analysis of the image twin system and the 3T6 body scanner. J Text Apparel Technol Manag 1(2)

2. ISO 20685 (2005) 3D scanning methodologies for internationally compatible anthropometric. ISO Standard 20685, International Standard Organisation, Geneva
3. ISO 20685 (2015) 3D scanning method for internationally compatible anthropometric databases. ISO Standard 20685, International Standard Organisation, Geneva
4. Braganc AS, Arezes P, Carvalho M, Ashdown S, Xu B, Castellucci I (2017) Validation study of a kinect based body imaging system. Work 57:9–21
5. Han H, Nam Y, Choi K (2010) Comparative analysis of 3D body scan measurements and manual measurements of size Korea adult females. Int J Ind Ergon 40:530–540
6. Kouchi M (2014) Anthropometric methods for apparel design: body measurement devices and techniques. In: Anthropometry, apparel sizing and design. Woodhead Publishing Limited in association with the Textile Institute, Cambridge, United Kingdom, pp 79–80
7. ISO 8559 (1989) Garment construction and anthropometric surveys—Body dimensions. ISO Standard 8559, International Organization for Standardization, Geneva
8. Vonk T, Danen H (2015) Validity and repeatability of the sizestream 3D scanner and Poikos modeling system. In : 6th International conference on 3D body scanning technologies, Lugano
9. Koepke N, Zwahlen M, Wells J, Bender N, Henneberg M, Rühli F, Staub K (2017) Comparison of 3D laser-based photonic scans and manual anthropometric measurements of body size and shape in a validation study of 123 young Swiss men. PeerJ 2980
10. Koo T, Li M (2016) A guideline of selecting and reporting intraclass correlation coefficients for reliability research. J Chiropract Med 15(2):155–163
11. Kouchi M, Mochimaru M (2011) Errors in landmarking and the evaluation of the accuracy of traditional and 3D anthropometry. Appl Ergonom 42:518–527
12. Kuehnapfel A, Ahnert P, Loeffler M, Broda A, Scholz M (2016) Reliability of 3D laser-based anthropometry and comparison with classical anthropometry. In: www.nature.com. https://www.nature.com/articles/srep26672 https://www.nature.com/articles/srep26672
13. Daniell N, Olds T, Tomkinson G (2012) Technical note: criterion validity of whole body surface area equations: a comparison using 3D laser scanning. Am J Phys Anthropol 148:148–155
14. Gordon CC, Blackwell CL, Bradtmiller B, Parham JL, Barrientos P, Paquette SP et al (2012) Anthropometric survey of U.S. Army personnel: Methods and Sumamry Statistics (ANSUR II). Comprehensive anthropometric survey of U.S. Army Soldiers (ANSUR II). Massachusetts: U.S. Army Natick Soldier Research, Development and Engineering Center Natick; 2012. Report No.: NATICK/TR-15/007
15. Simenko J, Cuk I (2016) Reliability and validity of NX-16 3D body scanner. Int J Morphol 34(4):1506–1514
16. Wang J, Gallagher D, Thornton JC, Yu W, Horlick M, Pi-Sunyer FX (2006) Validation of a 3-dimensional photonic scanner for the measurement of body volumes, dimensions, and percentage body fat. Am J Clin Nutr 83:809–816
17. Wells JCK, Stocks J, Bonner R, Raywood E, legg S, Lee S et al (2015) Acceptability, precision and accuracy of 3D photonic scanning for measurement of body shape in a multi-ethnic sample of children aged 5–11 years: the SLIC study. PLoS ONE 10(4)
18. Zancanaro C, Milanese C, Lovato C, Sandri M, Giachetti A (2015) Reliability of three-dimensional photonic scanner anthropometry performed by skilled and naïve operators. Int J Ergonom 5(5):1–11
19. Bougourd JP, Dekker L, Ross PG, Ward JP (2000) A Comparison of women's sizing by 3D electronic scanning and traditional anthropometry. J Text Inst 91(2):163–173
20. Paquette S, Brantley JD, Corner BD, Li P, Oliver T (2000) Automated extraction of anthropometric data from 3D images. In: Human factors and ergonomics society annual meeting proceedings; 2000: HFES, pp 727–730
21. Robinettea KM, Daanenb HAM (2006) Precision of the CAESAR scan-extracted measurements. Appl Ergonom 37:259–265
22. Weinberg SM, Scott NM, Neiswanger K, Brandon CA, Marazita ML (2004) Digital three-dimensional photogrammetry: evaluation of anthropometric precision and accuracy using a Genex 3D camera system. Cleft Palate Craniofac J 41(5):507–518

Design Construction and Performance Analysis of a Bobbinless Lockstitch Sewing Machine to Increase the Effectiveness in Industrial Production

Md. Nazmul Haque Nihad, Zihan Rana Zim, and Mahfuj Ul Sakik

1 Introduction

The Lockstitch sewing machine is widely used in textile industries for its efficacy and simplicity. This stitch is formed by two sets of threads which are needle thread and bobbin thread. The upper thread runs from a spool kept on a spindle on top of or next to the machine, using a mechanism that works on a principle of tensile force through the take-up arm, and finally through the hole in the needle. Meanwhile, the lower thread is wound onto a bobbin, which is inserted into a case in the lower section of the machine below the material. For a large amount of production, the size of the bobbin package is quite small and for which, after a certain time, it needs to be replaced repeatedly. This modification greatly simplifies the uninterrupted supplying of the bottom thread and qualifies the entire operation to be carried out completely automatically without replacement of any bobbin. The existing work is an object of the invention to provide a device capable of adapting a bobbinless coil of thread for use with the detecting apparatuses but the addition of a device makes the operating the machine a bit complex [1]. In previous work, the bottom thread was supplied directly from a bulk Source, such as a large cone-shaped winding, being metered out from the source, and it was then injected into the basket cavity of a lockstitch sewing machine (where the bobbin would normally be placed) to cooperate with a conventional reciprocating needle carrying a top thread to form lockstitches between the top thread and the bottom thread but the production of this design is full of extra machine parts which made the industrialization quite impossible [2].

Md. N. H. Nihad (✉) · Z. R. Zim
Department of Textile Engineering, National Institute of Textile Engineering and Research (NITER), Dhaka, Bangladesh
e-mail: nazmulnihad@gmail.com

M. U. Sakik
Department of Mechanical Engineering, National Institute of Textile Engineering and Research (NITER), Dhaka, Bangladesh

A. Majumdar et al. (eds.), *Functional Textiles and Clothing 2020*,
https://doi.org/10.1007/978-981-15-9376-5_3

2 Materials and Methodology

This work mainly focuses on the continuous bottom thread supply where stitch loop forms by using a half-circular rotation of "Rotary Hook/Loop Taker/Shuttle." The half rotations performed here by turning the middle bed shaft into a crankshaft on the right side were facing the machine pulley. There joined a connecting rod which is connected with a rack. The rack is being held inside a frame where it gets a reciprocating motion from the circular motion of the crankshaft through the connecting rod. As a result, the rack is proving a half-circular rotary motion to the pinion (1) adjoined on a shaft that is connected with another gear (2) on the left side and supported by a bush placed under the machine bed. The gear (2) is being provided with the same half-circular motion from the pinion (1) directly (Fig. 1).

Gear (2) is meshing with an idle gear (3) which rotates on a shaft attached to a bush with the bed. The idle gear (3) drives the spur gear (4) which is attached with the loop taker. The Loop Taker gets the rotation and being hold by a cylindrical pipe which is fastening to the body. This pipe is the ultimate bottom thread path where one end used for thread insertion (5) and the other end exposed to the inside of the Loop taker. The inside portion is divided into two sections; one is for loop rotation and the other one for the bottom thread supply. The loop rotation part is modified (Fig. 3) into a nearly half-circle shaped bracket in such a manner that the upper thread can pass smoothly. The other section is made of another half-circle shaped bracket with a tensioning device connected to the opening of that cylindrical pipe. During the stitch operation, the loop taker gets the upper thread from the needle and rotating half around and then releases it in the midway. But the upward movement of the needle creates a tensile force that takes the loop up through the gap between the two half-circular brackets. On that spot, the bottom thread passing through the

Fig. 1 The rack and pinion

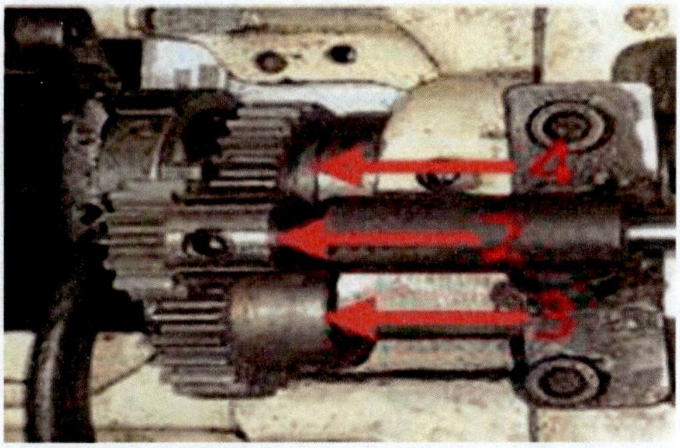

Fig. 2 Gearing mechanism

Fig. 3 The modified hook with gear (4)

tensioning device remains perpendicularly in the way of the loop and for the motion, it gets trapped which consequently creates the stitch on the fabric. Continuously, the stitch keeps forming properly by both the upper and bottom thread (Fig. 2).

The following Fig. (4) represents our designed parts assembled on the machine.

As mentioned above, due to the continuous flow of sewing thread and motion of the needle there is a fact about an increase of temperature that gives rise to the risk of sewing operation. There is no sewing operation can be found that doesn't require any kind of stoppage. For various kinds of designs and sewing requirements, there always remain small breaks in turning points of the seams. Alongside that advantage for ensuring the highest abrasion resistance, Glace Cotton threads were used in this operation [3]. For both upper and bottom thread, the same kind of threads was used cause like the upper thread on the way to the needle, a number of guides or incurved thread paths are present here in the bobbin less stitching mechanism. Furthermore, it was observed in previous studies that sewing speed is the most important factor for needle temperature followed by sewing time, number of layers and the stitch density [4]. None of these factors were the objective of this work rather the main focus was on the construction of the design and its analysis of performance to produce stitch

Fig. 4 The accomplished construction of the design

without the help of the bobbin. The sample fabric used here also comprised of 80 % Cotton and 20 % Polyester. The fabric just had the function of expressing output whether the stitch was forming properly or not for which it didn't require any special properties. Conventional sewing needle made of platinum and titanium alloy were used to ensure corrosion and abrasion resistance. The installed lubrication system was modified by adding more lubricating points on the modified parts of the design. All the Gears, Shafts, and other parts were used in this modification was made of stainless steel with high quality assuring sustainability in a continuous motion (Figs. 4 and 5).

3 Results and Discussion

The following graphical analysis signifies that just by using the modified bobbinless Lock Stitch Sewing machine a remarkable amount of production time is being accumulated. It was observed that for a skilled operator it takes about 30 s to replace an empty bobbin and for continuous production, in each hour the operation is done twice. To sum up, a sewing line with 10 Lockstitch sewing machine consumes 10 min every hour which results in about 1.6 h throughout the whole shift. Ultimately, the increase in efficiency is about 16.66 % more than the conventional lockstitch machine and the amount increases exponentially with a more large number of machines used in a continuously running garments factory. This numerical analysis is done based on the lockstitch sewing machines that have been modified. So, it can be claimed along

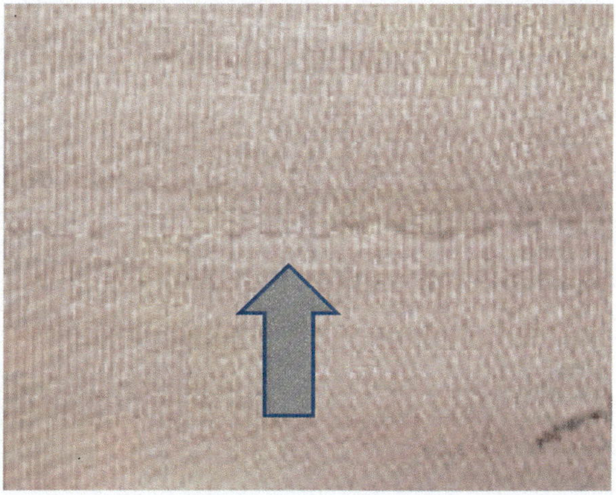

Fig. 5 The stitched sample

with other unwanted technical difficulties this modification can increase a notable efficiency in seam production.

In the graph above, the relation between both of the operations has been shown. The X-axis showed the total shift time and the Y-axis showed the effective time of operation in the graph. It has been seen that for the bobbin less operation, the lockstitch sewing machine can run efficiently almost all the shifting time, neglecting any intended break. But with the conventional operation, it took around 80 min in 8 h just to change the bobbin case. So there has been a distinguishing result between these two operations.

The designed machine overcomes the time consumption problem promisingly for its direct thread path from an outer package. Besides, other problems like in conventional lockstitch sewing machines when a bobbin runs out in the middle of a continuous seam overlapping occur to continue the production. Sometime entanglement occurs which is also become eliminated by this constructed lockstitch sewing machine. The performance analysis also shows some better characteristics with a little flaw. The motion derived from the machine pulley now turned into a half rotation through rack and pinion but for using the same specification for gears like diameter, teeth no. between identical parts, the ultimate motion remains equivalent. However, for the bottom thread path, the cylindrical pipe and the modified hook ensures the optimum tension for regulating proper stitch. Furthermore, maintenance is quite easier for keeping the expected efficiency. Stitch class: 301 is produced here using a needle and bottom thread but without using any bobbin. The strength of the stitch compared to conventional lockstitch sewing machine no detectable difference appeared. Even, by using high strength heavy fabrics can be sewed spontaneously (Fig. 6).

The costing is quite reasonable for this modification. The aforementioned details express that the difference between a conventional lock stitch sewing machine and the modified one is a small number of shafts, gears and some actions such as grinding,

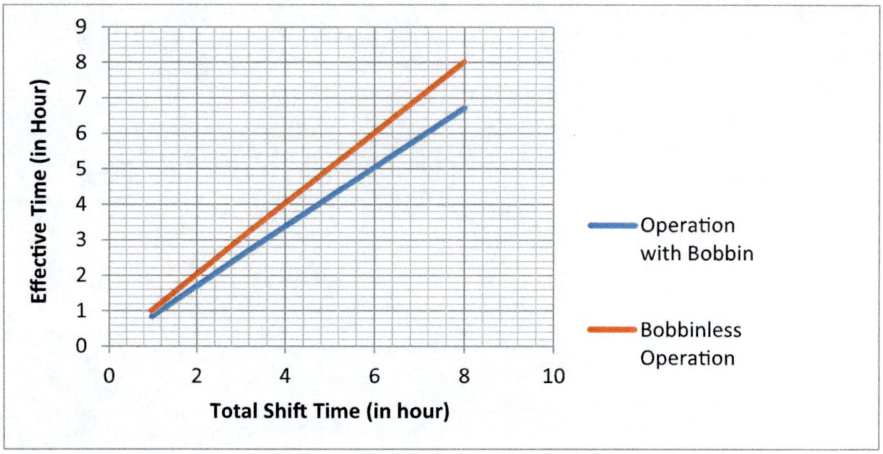

Fig. 6 The graph of effectiveness in an 8 h shift

cutting, etc., for yielding the modification. So, the possible increase in the cost of this modification is approximately 200 USD (17000 BDT). This increase is about only 30 % cost of any modern conventional lock stitch sewing machine. This is very promising to get such effective modification at a cheap cost.

4 Conclusion

The constructed design is simple enough to industrialize at a reasonable cost. Moreover, no additional installation is needed and the previous setup for pre-wound bobbin can be neglected. It's quite effective in terms of a large number of productions. The objective of the work is achieved by developing such an efficient design which opens up a new field for further research. With more support and investment bulk production would be possible to spread out the impact of this research. In the RMG sector, such modification and analytical study are necessary for export-oriented countries like Bangladesh, India, Vietnam, China, etc. The conducted modification is applicable for any kind of conventional single needle Lockstitch sewing machine with the bobbin and rotary hook. Because the modification actually reshaped the bobbin package holding chamber and the stitching mechanism with a newly modified hook. There are some limitations to the performance analysis with the need for further amendment. The indicated construction can be approached to advance the functional designing of specific elements.

References

1. Mardix et al (1993) Device for Use with a Bobbinless. United States Patent (US5570646A), pp 1–2
2. Herman Rovin (1970) Lockstitch sewing method and system providing bobbinless feed of the bottom thread from a bulk source. United States Patent (US5570646A), pp 1–2
3. Textile Apex Blog (2020) https://textileapex.blogspot.com/2018/12/sewing-thread-definition-type-uses.html. Accessed 06 Aug 2020
4. Mazari A, Zhu G, Havelka A (2012) Sewing needle temperature of an industrial lockstitch machine. Adv Mater Res 627:456–460

Moisture Management Properties of Ring Vis-à-vis Rotor Yarn Plated Knit Structures

Yamini Jhanji, Deepti Gupta, and V. K. Kothari

1 Introduction

Heat and mass transmission from the human body to the environment or vice-versa takes place through clothing which serves as a second skin. Thermo-physiological properties of textiles are related to the thermal properties, air permeability and moisture management properties along with the drying ability of the textiles. Moisture management properties of textiles are influenced by several fibres, yarn and fabric constructional variables. Plating is an innovative knitted fabric production technique to obtain bi-layered fabrics. Plated knit structures are characterized by distinct yet integrated inner and outer layers. The flexibility in the selection of contrastingly different fibre and yarn constituents in the two layers of plated fabrics make them suitable and versatile for applications like intimate wears, sportswear, active wear etc. The main elements of two-layered plated knitted structures are:

Inner (next to skin) layer: This layer is in direct skin contact and consists of a conductive and diffusive hydrophobic component which helps in removal and transportation of sweat to the outer layer.

Outer (exposed to environment) layer: This layer is in contact with the environment and consists of absorptive hydrophilic components which provide a large area for sweat absorption and evaporation to the outside environment [1, 2]. Fibre parameters, process parameters and yarn production technologies are reported to change the yarn structure and the associated fabric properties. Ring and rotor spun yarns vary widely in their structures which contribute to the entirely different properties of the two yarns. Ring-spun yarn has an ideal cylindrical helical structure with the same number of turns per unit length in each helix, uniform specific volume and maximum

Y. Jhanji (✉)
Technological Institute of Textile and Sciences, Bhiwani 127021, India
e-mail: yjhanji@gmail.com

D. Gupta · V. K. Kothari
Indian Institute of Technology Delhi, New Delhi 110016, India

A. Majumdar et al. (eds.), *Functional Textiles and Clothing 2020*,
https://doi.org/10.1007/978-981-15-9376-5_4

packing density in the outermost zone of the yarn cross-section. Rotor spun yarn has a bipartite structure with an inner core which forms the bulk of the yarn and an outer zone of wrapper fibres occurring irregularly along the core length. Rotor yarn shows maximum packing density in the first zone from the core. The core part of rotor yarn is relatively dense structure; sheath part is a less dense structure with belly-bands [3–5]. Several researchers have studied the influence of various yarn variables such as yarn types, yarn spinning systems and fibre variables such as fibre fineness, fibre cross-sectional shapes on comfort properties of woven and knitted fabrics. Behera et al. [6] compared the comfort properties of the ring, rotor and friction spun yarn fabrics and suggested that in the absence of perspiration, rotor spun yarn would be superior to ring-spun yarns. Lord [7] investigated the relative moisture uptake characteristics of ring and open-end yarns and reported that open-end yarn wicked better and more evenly but for the given yarn count elevated about the same volume of water as ring yarn. Erdumlu and Saricam [8] studied the wicking and drying properties of vortex spun yarns and knitted fabrics in comparison with ring-spun yarns and fabrics and concluded that fabrics knitted from ring-spun yarns wicked and absorbed water more evenly than fabrics knitted from vortex spun yarns. There is a lack of systematic study on the influence of yarn spinning systems on thermo-physiological properties of plated knit structures. Yarns made on different spinning systems vary in their structure and packing which may influence the transfer of moisture vapour and liquid moisture through fabrics. Therefore, the selected variable is crucial in influencing the moisture management properties of fabrics. The present study, therefore, attempts to study moisture management properties of ring vis a vis rotor yarn plated knit structures.

2 Materials and Methods

2.1 Materials

Cotton and polyester fibres were used in the production of the yarn samples. Cotton carded roving of 0.9 hank was used to spin 24 Ne single ring-spun yarn on LR G5/1 ring spinning system. Cotton carded sliver of 0.12 hank was used for the production of 24 Ne single rotor yarn on Trytex rotor spinning machine. The polyester fibre of 1.1, 3.3 decitex and circular profile was used to spin 24 Ne single ring-spun polyester yarns on 6/S LMW pilot plant ring frame. The SEM images of the ring and rotor spun yarn is provided in Fig. 1. The three yarn samples in totality were used for the preparation of four single jersey plated fabrics. All the samples were prepared in plating relationship with ring and rotor cotton yarn in the outer and polyester fibres in the inner layer. The fabric samples were prepared on hand-operated flat-bed knitting machine (Elex, China) with machine gauge of 14, needle bed of 42 inches and 588 needles in each bed. Table 1 summarises the plated fabric details.

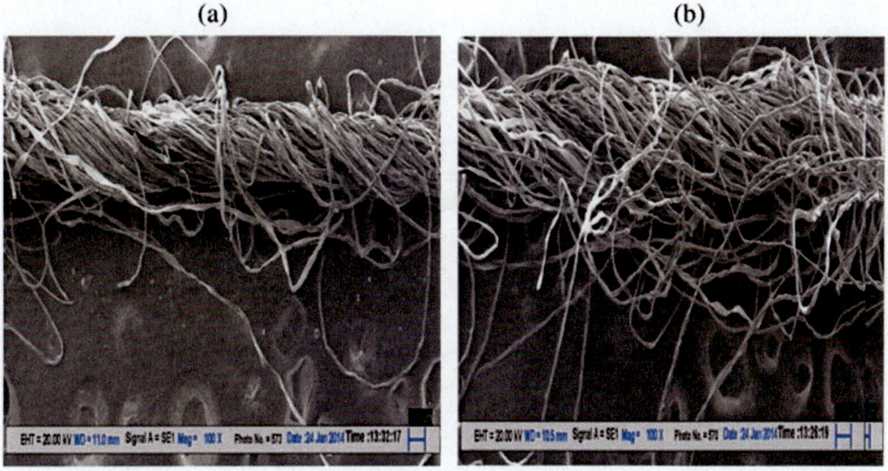

Fig. 1 SEM images of **a** Ring yarn structure, **b** Rotor yarn structure

Table 1 Details of developed plated fabrics

Sample Code	PET fibre linear density (dtex)	Yarn spinning system
$PCR_{1.1}$	1.1	Ring
$PCRO_{1.1}$	1.1	Rotor
$PCR_{3.3}$	3.3	Ring
$PCRO_{3.3}$	3.3	Rotor
$PCR_{1.1}$	1.1	Ring

Outer layer-C cotton, *Inner layer-P* polyester, *R* ring, *RO* rotor, *1.1, 3.3* polyester fibre dtex

2.2 Methods

The thickness of fabrics was measured on Essdiel thickness gauge at a pressure of 20 gf/cm^2. The areal density of samples was determined according to ASTM D-1059. Moisture vapour transmission rate of the fabrics was tested on moisture vapour transmission cell (MVTR cell) (Grace, Cryov ac division). The yarn structures were determined using the scanning electron microscope. Absorbent capacity and trans planar wicking of test samples were determined by Gravimetric absorbency tester. Moisture management tester (MMT) (SDL Atlas, Hong Kong) (AATCC Test method 195-2009) was used for testing the liquid moisture transfer properties of the test fabrics (Fig. 2).

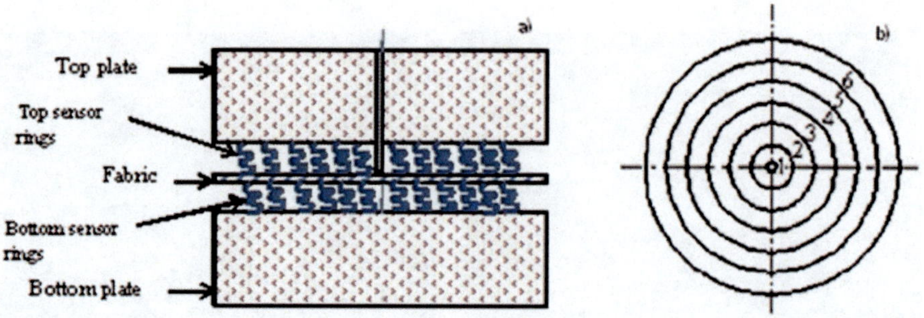

Fig. 2 Sketch of MMT sensors **a** sensor structure, **b** measuring rings

Table 2 Physical properties and moisture vapour transmission rate of plated knitted fabrics

Sample code	Thickness (mm)	Areal density (g/m^2)	Porosity (%)	Moisture vapour transmission rate (g/m^2/24 h)
$PCR_{1.1}$	0.927	248	81.68	7.25
$PCRO_{1.1}$	0.951	255	81.63	6.31
$PCR_{3.3}$	0.974	238	83.46	9.46
$PCRO_{3.3}$	0.985	250	82.56	8.81

3 Results and Discussion

3.1 Moisture Vapour Transmission Rate

Ring yarn fabrics exhibited a higher rate of moisture vapour transmission (7.25 to 9.46 g/m^2/24 h) as compared to the rotor yarn fabric samples (6.31 to 8.81 g/m^2/24 h) which may be attributed to the high porosity of ring yarn fabrics (Table 2). Free open spaces in the fabric facilitate easy passage of moisture vapour owing to the diffusion of moisture vapour through the air. As diffusion through fibrous material is restricted by fibre diffusivity, hence ring yarn fabrics with higher air volume fraction were found to be more permeable to moisture vapour transmission.

3.2 Trans Planar Wicking

Trans planar wicking of ring yarn fabrics was found to be higher as compared to their rotor yarn counterparts (Table 3). This may be attributed to the difference in yarn structure of two yarns used in the study. Better fibre alignment, a higher degree of compactness due to high packing density might have favoured the formation of a large number of continuous and small diameter capillaries in ring yarn fabrics. Rotor yarn

Table 3 Water uptake in Trans planar wicking (g)

Sample code	Time (s)									
	10	20	30	40	50	60	70	80	90	100
$PCR_{1.1}$	0.66	1.28	1.81	2.26	2.58	2.79	2.90	2.95	2.98	3.02
$PCRO_{1.1}$	0.004	0.12	0.54	1.16	1.74	2.21	2.50	2.65	2.71	2.73
$PCR_{3.3}$	0.30	1.12	2.08	2.84	3.27	3.42	3.49	3.55	3.59	3.63
$PCRO_{3.3}$	0.29	1.19	1.93	2.53	2.78	2.95	3.07	3.21	3.28	3.33

on the contrary displays randomness in the internal structure with dense core, less dense sheath and belly bands. The continuity of capillaries may be disturbed by tight wrappings along yarn length. In the light of above facts, it can be argued that randomness and tight wrapping along yarn length and more open structure compared to their ring yarn counterpart may disrupt the continuity of capillaries thereby inhibiting the liquid movement through capillary wicking in rotor yarn fabrics.

3.3 *Absorbent Capacity*

Water absorption of the textile fabrics is a crucial property in wearer comfort as it determines the liquid sweat holding capacity of the fabrics. Rotor yarn fabrics showed higher absorbent capacity as compared to ring yarn fabrics over a test period of 100 s as shown in Fig. 3. The observed trend may be attributed to high thickness,

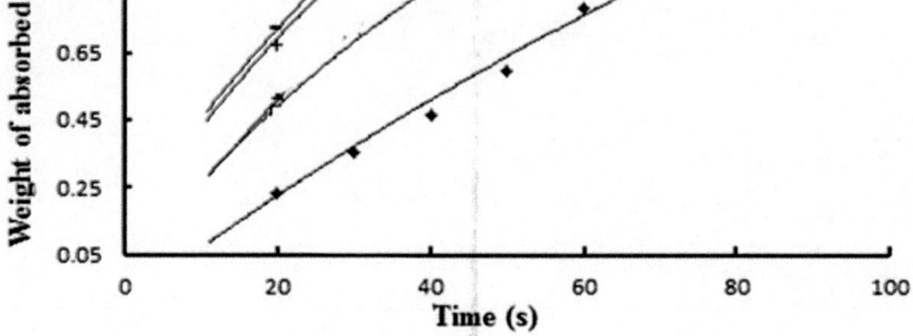

Fig. 3 Absorbent capacity of plated knitted fabrics

areal density and bulk density and hence more water absorption by rotor yarn fabrics as compared to their ring yarn counterparts.

3.4 Moisture Management Properties

Table 4 shows the moisture management indices of test samples. It was observed that top (inner layer) and bottom (outer layer) wetting time was higher for rotor yarn fabrics compared to their ring yarn counterparts suggesting that former would take longer to get wet on initial exposure to test liquid. Figures 4 and 5 show the water content curves for ring and rotor yarn fabrics. Spreading speed and one-way transport capacity which indicates the effectiveness of fabric in liquid spreading and transporting from inner to the outer layer was higher for ring yarn fabrics compared to rotor yarn fabrics. It can, therefore, be concluded that ring yarn fabrics would result in better spreading of test liquid in both inner and outer layers (higher SSt & SSb) and would be more effective in liquid transfer from top (inner/next to skin layer) to bottom (outer) layer as suggested by higher one-way transport capacity.

Table 4 Moisture management indices of plated knitted fabrics

Sample code	WTt (s)	WTb (s)	SSt (mm/s)	SSb (mm/s)	OWTC
$PCR_{1.1}$	2.91	2.06	2.38	3.44	622.57
$PCRO_{1.1}$	6.75	2.25	1.47	2.56	483.75
$PCR_{3.3}$	3.66	5.16	2.22	2.10	573.32
$PCRO_{3.3}$	7.45	6.23	1.25	1.83	428.62

WTt Top wetting time, *WTb* Bottom wetting time, *SSt* Top spreading speed, *SSb* Bottom spreading speed, *OWTC* One-way transport capacity

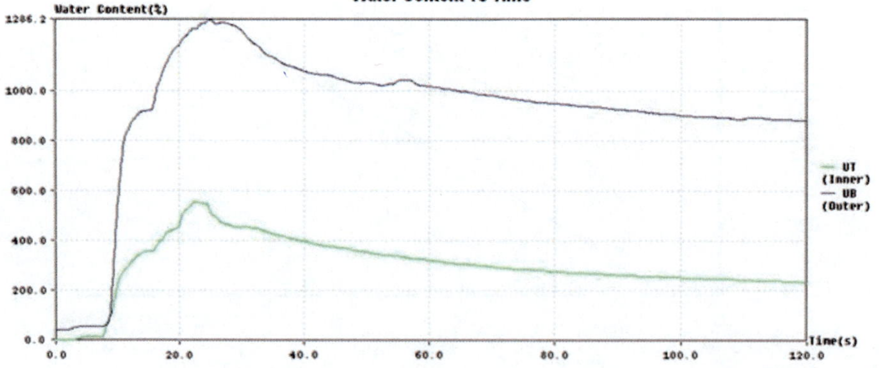

Fig. 4 Water content versus time curve for inner (top) & outer (bottom) layers of ring yarn fabrics

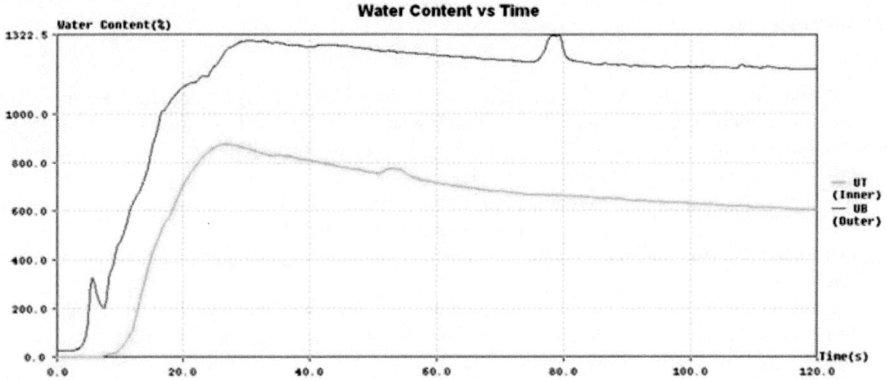

Fig. 5 Water content versus time curve for inner (top) & outer (bottom) layers of rotor yarn fabrics

The test results suggested that ring yarn fabrics owing to higher moisture vapour transmission rate, trans planar wicking, lower wetting time, higher spreading speed, and one-way transport capacity performed better in liquid moisture transmission as compared to rotor yarn fabrics. Liquid moisture transmission is primarily affected by capillary size and geometry. More continuous capillaries of smaller diameter as a result of uniform helical structure and lower diameter of ring yarn unlike rotor yarn resulted in better moisture management properties of the former. However, the absorbent capacity of fabrics is related to the mass of water initially absorbed by fabrics which in turn depend on fabric's areal density. The observed trend, that is the higher absorbent capacity of rotor yarn fabrics can be explained by their higher areal density compared to ring yarn counterparts.

4 Conclusions

The study proposes the development of plated knit structures with the outer layer composed of two different yarn structures and the comparison of developed fabrics for their moisture management properties. Ring yarn fabrics exhibited higher moisture vapour transmission rate, trans planar wicking, lower wetting time, higher spreading speed and one-way transport capacity as compared to rotor yarn fabrics suggesting that ring yarn fabrics would result in better spreading of test liquid in both inner and outer layers and would be more effective in liquid transfer from the top (inner/next to skin layer) to bottom (outer) layer. The findings of the study help us to conclude that plated knit structures with ring yarn in the outer layer are suitable choice where body need to dissipate sweat both in vapour and liquid form with respect to fabrics using rotor spun cotton yarn in the outer layer, as the latter show higher absorbent capacity and would be slow drying with poor one-way transport capacity.

References

1. Geraldes M, Lubos H, Araujao M, Belino NJR, Nunes MF (2008) Engineering design of the thermal properties in smart and adaptive knitting structures. Autex Res J 8:30–33
2. Bivainyte A, Mikucioniene D (2011) Air and water vapour permeability of double-layered weft knitted fabrics. Fibres Text Eastern Europe 19:64–68
3. Kumar RC, Kumar AP, Senthilnathan P, Jeevith R, Anbumani N (2008) Comparative studies on ring, rotor and vortex yarn knitted fabrics. Autex Res J 8:100–105
4. Ozturk MJ, Nergis B, Candan C (2011) A study of wicking properties of cotton-acrylic yarns & knitted fabrics. Text Res J 81:324–328
5. Basu A (2009) Yarn-structure properties relationship. Indian J Fibre Text Res 34:287–289
6. Behera BK, Ishtiaque SM, Chand S (1997) Comfort properties of fabrics woven from ring, rotor and friction spun yarns. J Text Inst 1:255–259
7. Lord PR (1974) A comparison of the performance of open end and ring spun yarns in terry toweling. Text Res J 44:516–518
8. Erdumlu N, Saricam C (2013) Wicking and drying properties of conventional ring and vortex spun cotton yarns and fabrics. J Text Inst 104:1284–1287

Functional and Protective Clothing

Exploring the Need for Functional Clothing to Optimise Metabolic Consumption

Lindsay D'Arcy, Mike Fray, and Jo Barnes

1 Introduction

To maintain a stable body weight, energy intake must equal energy expenditure, a state known as energy balance [1]. However, a wealth of knowledge supports the notion that physical activity has significantly declined over the past century [2]. Sedentary occupations, the rise of technology and motorised transportation have made life more convenient than ever and thus, the drive for saving time, has consequently decreased the metabolic energy requirements to overcome every day, non-exercise activities [3]. Large numbers of people are expending less than they consume and consequently, increasing rates of obesity reflect this lack of energy balance [1]. It is more difficult to maintain an energy balance in the modern environment where every day, voluntary and spontaneous lifestyle physical activity is not a given. Intervention is therefore required to reclaim the energy expenditures that were once a given through every day, physically active lives. Purposeful exercise is widely adopted as a method to increase energy expenditure, however, barriers to exercise are experienced by many and typically depend on age, health conditions, body mass index, etc. Younger adults generally cite lack of time as their main constraint to exercise [4], and elderly people most frequently cite poor health as their leading barrier, such as fear of injury or falling [5]. Overweight women generally cite feeling too overweight to exercise [6], feeling body conscious [7] and lacking the confidence to exercise in the gym environment [8], thus overweight adults exhibit lower levels of adherence to exercise than normal-weight adults [9]. Nevertheless, in recent years, researchers have stressed the importance of non-exercise activity thermogenesis (NEAT) contributions to the energy balance and physiology weight control [3]. The extensive list of activities contributing to NEAT include occupational tasks, sitting, standing, walking, talking,

L. D'Arcy (✉) · M. Fray · J. Barnes
Loughborough University, Loughborough L11 3TU, UK
e-mail: l.darcy@lboro.ac.uk

A. Majumdar et al. (eds.), *Functional Textiles and Clothing 2020*,
https://doi.org/10.1007/978-981-15-9376-5_5

fidgeting etc., [3] however, what remains unexplored as a contributor to NEAT, is the influence of clothing on energy expenditure.

In the field of personal protective clothing, literature has evidenced that clothing can increase metabolic energy consumption using multiple methods including clothing weight [10, 11], clothing bulk [11, 12], multiple layers [13–15], high friction fabrics [16], fabric stiffness [11, 12] and thermal/insulative properties [17]. Increases of 10–20 % in energy expenditure are not uncommon [10]. However, the contributions are limited regarding the application of these methods for purposefully increasing metabolic activity. The metabolic influences of clothing are considered design defects, given the field they are tested in. The drive for functional clothing development targets decreasing metabolic energy consumption, making clothing less resistance to movement [18].

The historical changes that have occurred in clothing norms are extreme. In the 1880s, women's clothing ensembles consisted of up to five layers, collectively weighing approximately 4.4 kg, today a contemporary woman's outfit averages at two layers, weighing 1.1 kg [19]. The development of lightweight fabrics and surge of athleisure trends have theoretically, reduced the energy demanded to move and thus, the influence of clothing on metabolic energy consumption could very well be at its lowest.

Resistance clothing has been adopted by athletes and exercise enthusiasts for training and conditioning purposes. The clothing enhances the energetic efforts required to perform physical activities during training and thus, the body builds strength in ways that will allow it to be faster or stronger once the resistance is removed [18]. Resistance is inflicted when an external force causes the skeletal muscles to contract and therefore work harder to overcome a task [20]. By increasing the wearable resistance, a person can achieve a range of performance-enhancing benefits [21]. It has been evidenced that wearable resistance may be achieved by various means including increasing aerodynamic drag [22, 23], weight resistance [24] and stretch resistance [25].

Considering a high population of women are overweight [26], have intentions to lose weight, yet experience multiple barriers to exercise [9], it proves worthwhile exploring if women might be interested in clothing that increases the metabolic energy cost required to perform everyday tasks and/or maximise exercise currently undertaken.

2 Aim of the Study

The overall aim of the study was to investigate women's approach to exercise and diet and to explore the potential interest of functional clothing to increase metabolic consumption.

3 Method

To reach a broad population of adult women and have them disclose weight management information comfortably and anonymously, a self-administered online survey was deemed suitable for the method of data collection. A self-administered questionnaire was developed and implemented using Jisc Online Surveys (formerly Bristol Online Survey – BOS) and was available via a web-link. The questionnaire was inclusively designed, to allow both exercisers and non-exercisers to answer the questions. It was limited to 20 questions and was designed to be completed in 5–10 min to maximise the response rate. Multiple choice questions allowed respondents to self-report weight and exercise management, whilst Likert rating scale questions allowed respondents to comment on their level of belief and preference [27]. An optional, open-ended question was also included for respondents to express any further views or opinions on the clothing.

3.1 Sampling Method

The sampling methods of the study were volunteer and snowball [28]. The sample was limited to adult women. No further inclusion criteria were adopted due to the scoping nature of the study. The initial sampling strategy was controlled at three main levels (1) friends/family/colleagues (2) social media public sharing and (3) weight management programmes. However, permission to recruit participants via weight management programmes was rejected as the study objectives did not fit with their current research strategies. To ensure this sample was identified in the analysis, the questionnaire allowed for participants to identify as a member of a weight management group. The sampling strategy was revised and dissemination of the survey was controlled at two top levels (1) friends, family and colleagues and (2) social media. The researcher contacted friends, family and colleagues to ask them to participate in the study (volunteer method). On completion of the questionnaire, participants were asked to inform other women who might be interested in completing the survey (snowball method). A further internal dissemination to colleagues and students at Loughborough University was achieved by submitting to the Loughborough University staff and student online noticeboard. The researcher also publicly advertised across social media channels, including Facebook, Instagram and LinkedIn. Follow up reminders were implemented after three weeks to allow for the final week of responses. The total duration of dissemination was 4 weeks.

3.2 Question Development

The aims of the study were outlined to identify the information required, which helped develop the types of questions to be included. The questionnaire was split into seven sections, including weight management, dietary habits, physical activity, occupation, workwear, functional clothing and personal information. The questionnaire used terminology that would be inclusively understood by a wide population, for example 'metabolic energy consumption' was referred to as 'burning calories'.

1. Weight management. The first section was dedicated to weight management. The questions were developed to understand women's current weight motivations; whether they were trying to lose weight, maintain weight, gain weight, or not managing their weight at all.

2. Dietary habits. The dietary habits section aimed to understand how women managed their diet. Respondents were asked if they were aware of their calorie intake and if so, how many calories they consumed per day (kcal/d), if they felt they consumed more calories than they expended and how they managed their diet. Diet management responses included a range of dietary regimens from calorie focus to intermittent fasting. This section also allowed for identifying those who were members of established weight management programmes.

3. Physical activity. Physical activity was introduced and defined in Sect. 3 as "any bodily movement produced by the skeletal muscles that uses energy" [29] and was broken down into three types of activity including (1) exercise: any type of physical activity that you engage in during your leisure time that is planned, structured, repetitive and intentional [29], (2) work-related activity: the physical activity you engage in during working hours, in the workplace and (3) commuting: the physical activity you engage in whilst travelling to and from work/education. The types of physical activity were explained to participants before self-reporting their activity levels to avoid misinterpretation. For example, a respondent may submit walking to work as exercise instead of commuting physical activity. By splitting the types of physical activity, the researcher could gain further insight into the environment the respondent was active in. Respondents were then asked if they engaged in any exercise, as defined, with non-exercisers skipping this section.

Exercisers were asked to self-report their exercise mode, frequency and duration. Exercise mode was used to determine the types of activities and movements undertaken by the participant. Respondents chose up to 3 of the 47 activities listed. Exercise frequency was measured against times per week and included options from once to 14 times per week, to allow for those who may exercise twice per day. The duration was measured against each session and included 15 min variables from < 15 min to > 2 h. Respondents were also asked if they tracked their physical activity and their method of tracking to understand whether the quantifying activity was important to them.

Participants were asked to rank their motivations and barriers to exercise using a five-point Likert scale from strongly disagree to strongly agree. The questions were inclusively designed for both exercisers and non-exercisers, to gather insight into what currently motivates exercisers and what would motivate non-exercisers to exercise. Exercise motivations included a 17 variable edit from the 51-item Exercise Motivations Inventory-2 (EMI-2) [30] which covered both intrinsic (e.g. enjoyment) and extrinsic (e.g. weight management) types of motivation [31]. Participants were asked to select their top three motivations. Perceived barriers to exercise covered five themes including time, energy, self-confidence, self-efficacy and health ability.

Occupational activity was self-reported and assessed using a single question with three response options including (1) sedentary: seated most of the day, (2) moderate: seated and standing and (3) active: rarely seated.

Commuting physical activity was self-reported and assessed against commuting modes including bicycle, car, public transport and walking. The duration spent commuting to and from the workplace was measured in 15 min variables from < 15 min to > 5 h. The question design allowed for multiple types of commuting for those who, for example, walked and used public transport. The question also allowed for those to report on working from home or if their current status was unemployed or retired and thus, did not commute to a place of work.

4. Occupation. Respondents were asked to identify their occupation type. Occupation type was important to understand the type of environment the respondent worked in, along with the types of movements, tasks and activities undertaken during working hours.

5. Workwear. Respondents were asked to describe their workwear across three variables including (1) no uniform - I wear whatever I choose to, (2) relaxed dress code – I am advised to dress appropriately and (3) strict uniform – I wear a uniform provided by my workplace. The question aimed to understand the occupational uniform principles respondents adhered to and the level of freedom they have with their clothing choices. Respondents were also asked the duration of workwear clothes worn, to determine if the clothing was worn outside of working hours during leisure time.

6. Functional clothing. Sect. 6 briefly educated respondents on how clothing can increase metabolic energy consumption and the potential benefits the clothing could support including weight management and muscle toning. The questions were paramount to the scoping study, to understand if respondents would like to wear the clothing, what would motivate them to wear the clothing (weight management, muscle toning or both) and when they would prefer to wear the clothing (commuting, exercising or during working/studying hours). Preferences could then be analysed against self-reported physical activity to determine the feasibility.

7. Personal information. The final section was dedicated to collecting personal information including age, height and weight to allow for Body Mass Index (BMI) to be calculated as weight in kilograms divided by height in meters, squared (kg/m^2). The final, open-ended question asked respondents to share any further views and opinions on the functional clothing in a free-text format. The intention was to allow for participants to discuss additional topics, concerns or questions, that were not captured in the previous closed questions.

3.3 Piloting

Prior to launching, the questionnaire was reviewed by two academics experienced in questionnaire design, followed by a pilot to adult women (n = 20). The sample included two experienced, industry qualitative researchers, friends, family and students at Loughborough University. Two participants were observed and timed during the completion of the questionnaire. Observations provided insight into the potential misinterpretation of questions and how fluidly participants moved through the sections. All participants completed the questionnaire independently and provided feedback. In response to the feedback obtained, several changes were made to the final version.

3.4 Ethics

The study was approved by the Ethics Sub-Committee for Human Participants of Loughborough University on the 6 September 2019. Ethical approval included an ethical clearance checklist, participant information sheet, informed consent form and a risk assessment for the online questionnaire. Due to the questionnaire being hosted online, hard copy signed consent forms were not feasible. The questionnaire's introductory foreword titled 'Participant Information and Consent' explained the research aims, data protection, confidentiality and informed respondents that the questionnaire collected no identifiable data, responses were anonymised, unless participants agreed to further participation in further research, and data would be handled in strict confidence. After reading, participants could then agree or refuse participation in the study.

3.5 Data Analysis

All statistical analyses were conducted using SPSS version 24 (SPSS, Inc. Chicago, IL). Chi-square tests were used to measure the degree of association across 4 categorical variables; a value of $p < 0.05$ was considered statistically significant. Descriptive statistics were used as a visual means to describe the survey data. All nominal, ordinal and scalar data were numerically coded when imported into SPSS. For the final open-ended question, a thematic analysis was conducted to identify themes, which allowed for coding into nominal data. The value of metabolic equivalents (METs) was assigned for each physical activity listed according to a compendium of physical activities [32]. The MET values were not visible to respondents, however, the values assigned allowed for the nominal variable (exercise mode) to be translated into scalar data (MET) to allow for an understanding and comparison of physical activity intensity.

4 Results

4.1 Participants

The sample size of the questionnaire respondents was n = 502 female. Many respondents were aged 26–35 (n = 173). More exercisers (n = 414 82.5%) were surveyed than non-exercisers (n = 88 17.5%) (Fig. 1) and 77.7 % (n = 390) reported to be managing their weight for weight loss or maintenance purposes.

Approximately half of the participants were normal weight (n = 254 50.6 %) and 47.4 % (n = 238) were overweight or obese (Table 1).

Of the total women surveyed, 83.5 % (n = 419) were interested in wearing functional clothing to increase metabolic energy consumption. Of this group, 79.9 % (n = 335) claimed to be managing their weight for either weight loss or maintenance purposes, 82.1 % (n = 344) engaged in exercise and 72.8 % (n = 305) managed their diet. However, of those who reported to be managing their diet and engaging in exercise 93.1 % (n = 390) felt they still consumed more calories than they expended.

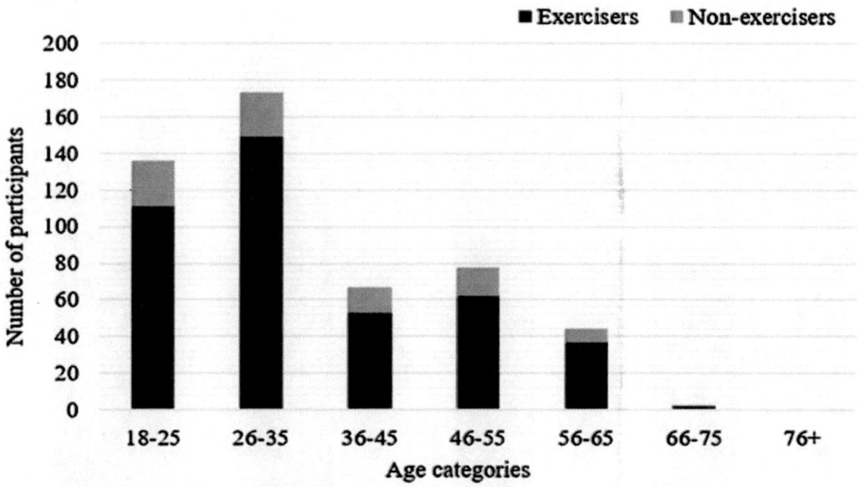

Fig. 1 Total participants surveyed by age category

Table 1 Total female participants surveyed with respect to their BMI status

Categorical variable	BMI status	BMI	(n (%)
1	Underweight	0 <= 18.5	10 (2 %)
2	Normal weight	18.5 <= 24.9	254 (50.6 %)
3	Overweight	24.9 <= 29.9	144 (28.7 %)
4	Obese	>30	94 (18.7 %)

The analyses revealed a significant association between BMI status and the interest in functional clothing to increase metabolic energy consumption ($p = 0.005$); as BMI increased, interest in the clothing also increased (Fig. 2). Obese women demonstrated the most interest in wearing functional clothing to increase metabolic energy consumption (91.5 %, n = 86), whilst underweight women were the least interested.

Respondents reported on what would motivate them, to wear the clothing. Of the women who were interested in wearing the clothing, all BMI groups felt that both weight management and muscle toning benefits would motivate them to wear the clothing. However, obese women demonstrated the most interest in solely weight management purposes, whilst underweight women reported that solely muscle toning benefits would motivate them. The analyses demonstrated that as BMI decreased, the level of interest in muscle toning increased.

Those who were not interested in the clothing (n = 83) were further analysed as a sample. The analyses demonstrated that 90.4 % (n = 75) of this group were dieters and exercisers and 66.2 % (n = 55) reported to be managing their weight for loss of maintenance purposes. Nevertheless, 50.6 % (n = 42) felt they consumed more calories than expended, which proposed further questions of the data, to understand why women would not be interested in wearing functional clothing if it aligned with their weight management motivations.

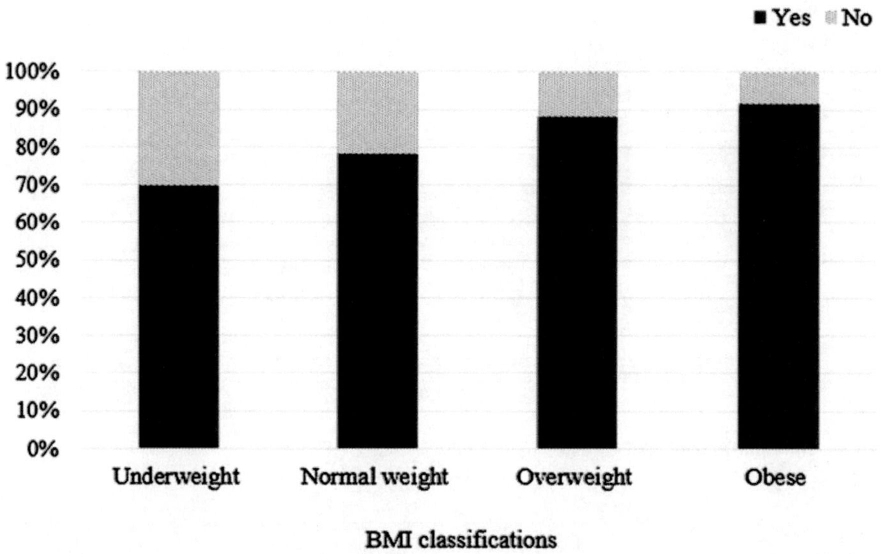

Fig. 2 Level of interest in functional clothing to increase metabolic consumption across BMI classifications

4.2 *Additional comments*

The thematic analysis provided further insight into women's views and opinions which presented themes of concern, regarding how the clothing might align with the wearer's lifestyle. Example quotes from participants are presented to support and provide context to the themes identified.

4.2.1 Aesthetics

> I think a big factor in wearing calorie expenditure clothing is if it looks fashionable and feels comfortable. If it's neither of those things then I wouldn't wear it.

(P435 – Overweight BMI. Non-exerciser, trying to lose weight)

4.2.2 Comfort

> I believe I would move less with clothing which feels awkward to move in.

(P329 – Obese BMI. exerciser, trying to lose weight)

4.2.3 Resistance Level

> Would prefer little changes throughout the entire day rather than a big difference over a short time e.g. commute.

(P268 – Overweight BMI. exerciser, trying to lose weight)

5 Discussion and Future Work

The purpose of this study was to explore the potential interest of functional clothing to optimise metabolic consumption. No previous research was found to explore this and thus, data could not be compared against previous findings. The sample size was large enough to present valuable insights of the women interested their motivations to wear the clothing and where they would prefer to wear it. Overall, there appears to be an apparent interest across all BMI categories, however, differences are presented regarding when women would prefer to wear the clothing and what would motivate them to wear it. This poses the challenge of delivering such clothes that would fit multiple needs and align with the motivations of all women interested. Follow up focus groups with participants will further explore how clothing might fit into the lifestyles of women; functionally, expressively and aesthetically, in accordance with the FEA consumer needs model [33]. Participants will be grouped according to their BMI, mutual interests and motivations to wear the functional clothing, to allow for common interests and concerns to be discussed.

6 Limitations

The scoping nature of the study aimed to explore general interest and opportunity of functional clothing to optimise metabolic energy consumption and therefore did not focus on the potential negative impacts of the clothing such as the potential discomfort of restricted movement. Whilst respondents were offered the opportunity to further comment on the clothing in the final open-ended question, the total number of valid responses (n = 61) did not reflect the concerns of the total sample of women surveyed (n = 502). Additionally, the questionnaire design did not include any visuals to reference the clothing under study which could have influenced the level of thorough understanding. However, this was purposeful, to allow respondents to speculate about the design, which will provide further insights in the secondary phase of the research. Another limitation, well-critiqued across literature, suggests that self-reporting physical activity leads to inaccurate results in certain BMI categories [34]. However, due to the scoping nature of the study, it was deemed suitable to deliver a general understanding of physical activity undertaken. Further insight into the user's physical activity and movement will be explored in the following stages of the research, which will adopt more in-depth methods including behavioural observation, motion sensors and wearable activity trackers.

7 Conclusion

The results determined a positive level of interest in functional clothing to optimise metabolic consumption across all BMI classifications and thus, the exploratory aim of the study has been achieved. In addition to this, the thematic analysis delivered further insight into the concerns and additional needs of women, which highlighted the importance of follow-up focus group discussions.

References

1. Hill JO, Wyatt HR, Peters JC (2013) The importance of energy balance. European Enforincol 9(2):111–115

2. Haskell WL, Lee IM, Pate RR, Powell KE, Blair SN (2007) Physical activity and public health: Updated recommendation for adults from the American College of Sports. Med Sci Sports Exerc 39(8):1423–1434
3. Levine JA (2004) Nonexercise activity thermogenesis (NEAT): environment and biology. Am J Physiol Endocrinol Metab 286(5):675–685
4. Daskapan A, Tuzun EH, Eker L (2006) Perceived barriers to physical activity in university students. J Sports Sci Med 5(4):615–620
5. O'Brien Cousins S (2000) "My heart couldn't take it": older women's beliefs about exercise benefits and risks. J of Gerontol: Psychol Sci 55:284–294
6. Ball K, Crawford D, Owen N (2000) Too fat to exercise? Obesity as a barrier to physical activity. Aust N Z J Public Health 24(3):331–333
7. Guess N (2012) A qualitative investigation of attitudes towards aerobic and resistance exercise amongst overweight and obese individuals. BMC Resh Notes 5(1):1–12
8. Arikawa AY, O'Dougherty M, Schmitz KH (2011) Adherence to a strength training intervention in adult women. J Phys Act Health 8(1):111–118
9. Ekkekakis P, Lind E (2006) Exercise does not feel the same when you are overweight: the impact of self-selected and imposed intensity on affect and exertion. Int J Obes 30(4):652–660
10. Dorman LE, Havenith G (2007) Effects of simulated clothing weight distribution on metabolic rate. European union project THERMPROTECT G6RD-CT-2002-00846 Report 2007-4
11. Duggan A (1988) Energy cost of stepping in protective clothing ensembles. Ergonomics 31(1):3–11
12. Dorman LE, Havenith G (2007) Examining the impact of protective clothing on range of movement. European Union project THERMPROTECT G6RD-CT-2002-00846, Report 2007-7
13. Amor AF, Vogel JA, Worsley DE (1973) The energy cost of wearing multilayer clothing. Army Personnel Research Establishment, Ministry of Defence. (Farnborough, Hants, UK). Report No. 18/73
14. Teitlebaum A, Goldman RF (1972) Increased energy cost with multiple clothing layers'. J Appl Physiol 32(6):8–9
15. Dorman LE, Havenith G (2007) Investigating the effect of clothing layers and their frictional properties on metabolic rate. European Union project THERMPROTECT G6RD-CT-2002-00846, Report 2007-5
16. Dorman LE, Havenith G (2007) Investigating the effect of wet layers and their frictional properties on metabolic rate. European Union project THERMPROTECT G6RD-CT-2002-00846, Report 2007-6
17. Parsons K (2003) Human thermal environments: The effects of hot, moderate and cold environments on human health, comfort and performance. Taylor and Francis, London
18. Watkins SM, Dunne LE (2015) Functional clothing design: from sportswear to spacesuits. Bloomsbury Publishing
19. Wells NM, Ashdown SP, Davies EH, Cowett FD, Yang Y (2007) Environment, design, and obesity. Environ Behav 39(1):6–33
20. Sundell J (2011) Resistance training is an effective tool against metabolic and frailty syndromes. Adv Prevent Med:1–7
21. Wescott WL (2012) Resistance training is medicine: effects of strength training on health.: The American College of Sports Medicine. Curr Sports Med Rep 11(4):209–216
22. Demarest N, MacDonald RC, Carbo JE (2020) Article of Apparel for Resistance Training. https://patents.google.com/patent/ US7856668B2/en?q=article&q=apparel &q = resistance &q=training&oq=article+of+apparel+for+resistance+training. Accessed 22 Jan 2020
23. Parkinson A (2020) Apparel item with integrated parachute structure. NIKE Inc. https://patents.google.com/patent/US10285463B2/en?oq=10%2c285%2c463. Accessed 24 Jan 2020
24. Macadam P, Cronin JB, Simperingham KD (2017) The effects of wearable resistance training on metabolic, kinematics and kinetic variables during walking, running, sprint running and jumping: A systematic review. Sports Med 47(5):887–906

25. Gottfried A (2020) Resistance shirt for muscle toning. https://patents.google.com/patent/US20160199684A1/en?oq=US+2016%2f0199684+A1. Accessed 24 Jan 2020
26. Butland B, Jebb S, Kopelman P, McPherson K, Thomas S, Madell J, Parry V (2007) Foresight. Tackling obesities: future choices – project report. Obesity reviews
27. Likert R (1932) A technique for the measurement of attitudes. Arch Psychol 22:140–155
28. Goodnan LA (1961) Snowball sampling. Ann Math Stat 32:148–170
29. Caspersen CJ, Powell KE, Christenson GM (1985) Physical activity, exercise, and physical fitness: Definitions and distinctions for health-related research. Public Health Rep 100(2):126–131
30. Markland D (2019) The exercise motivations inventory – 2 (EMI-2). In: Malthy J, Lewis CA, Hill AP (eds) Commissioned reviews of 250 psychological tests. Edwin Mellen Press, Lampeter, Wales, pp 58–60

32 Ryan RM, Deci EL (2000) Intrinsic and Extrinsic motivations: classic definitions and new directions. Contemp Educ Psychol 25:54–67

32. Ainsworth BE, Haskell WL, Herrmann SD, Meckes N, Bassett DR, Tudor-Locke C, Greer JL, Vezina J, Whitt-Glover MC, Leon AS (2011) Compendium of physical activities: classification of energy costs of human physical activities. A second update of codes and MET values. Med Sci Sports Exerc 43(8):1575–1581
33. Lamb JM, Kallal MJ (1992) A conceptual framework for apparel design. Cloth Text Res J 10(2):42–47
34. Sallis JF, Saelens BE (2000) Assessment of physical activity by self-report: status, limitations, and future directions. Res Q Exerc Sport 71(2):1–14

Development of a Unique Stab and Impact Resistant Material for Anti-riot Body Protector

M. S. Parmar, Neha Kapil, and Nidhi Sisodia

1 Introduction

In defense services mostly two types of body protectors are used prominently. These are (i) Bullet resistant jacket/vest and (ii) Anti-riot body protectors. A bullet-resistant jacket/vest is an item of personal armour that is worn on the torso. It helps to absorb or reduce the impact and penetration of bullet, fired from firearm. However, anti-riot body protector comprises helmet, neck protector, body vest, elbow protector, gloves, groin protector, thigh and leg guards. Its role is to protect the body from physical attacks—such as stone, knife, petrol bombs, stick, etc. These body protectors are made of hard shell (polycarbonate, polyamide, polypropylene, etc.) and foam (Polyvinyl chloride, Nitrile, etc.) and covered with specialized fabrics having fire, water and chemical (acid and Alkali) resistant properties. The demand for anti-riot body protector is increasing day by day amongst the police and paramilitary forces to cope up with riots like situation. Stabbing is one of the areas which needs to be addressed seriously. Security forces face a common threat of getting attacked by knife or other sharp-edged weapons. Briefly, stab threat can be classified into two categories as puncture and cut/slash. 'Puncture' refers to the penetration by instruments with sharp tips but no cutting edge, such as a spike or screw-drives, whereas 'cut' or 'slash' refers to contact with knives or scissors with continuous cutting edge. While designing body protectors, all types of stabbing probabilities are taken into account. These probabilities are based on scientific studies. In past, numbers of studies were carried out to analyze type and location of stab wounds. Some of the studies [1, 2] revealed that out of total 69 single fatal wounds cases, 50 were to the chest, 12 to the head & neck and 7 to the abdomen and lower limbs. In another study [3], out of 27 single fatal wounds, 18 were to the chest. Fligelstone [4] have carried out study on 92 cases mainly non-fatal injuries and observed that 37 resulted in

M. S. Parmar (✉) · N. Kapil · N. Sisodia
Northern India Textile Research Association, Sector-23, Rajnagar, Ghaziabad, India
e-mail: drmsparmar@nitratextile.org; msprmr@yahoo.com

A. Majumdar et al. (eds.), *Functional Textiles and Clothing 2020*,
https://doi.org/10.1007/978-981-15-9376-5_6

injury to the abdomen and 29 injuries to the chest. Researchers [5, 6] studied on the victims of stab injury in a hospital emergency admission and observed medical consequences of stabbing. On the basis of the study, they suggested requirement of suitable body armour for police personnel. In this study, wounds were analyzed using Injury Severity Score System. The Injury Severity Score (ISS) is an anatomical scoring system that provides an overall score for patients with multiple injuries. Each injury was assigned an Abbreviated Injury Scale (AIS) score and was allocated to one of six body regions (Head, Face, Chest, Abdomen, Extremities including Pelvis and External). Only the highest AIS score in each body region was used. The three most severely injured body regions have their score squared and added together to produce the ISS score. These studies revealed that the injury mechanism of stabbing was primarily blood loss and associated complications, or punctures to the chest leading to respiratory failure. All these studies revealed that fatalities from stabbing are in the majority of cases due to chest injuries. However, the saving of abdomen was also important. To understand the depth of penetration required in order to cause damage to vital organs and arteries by stabbing, a study [6] was carried out on 71 subjects. Under this whole body, computer tomography scan was used to determine skin to organ distances. It was observed that the distance of vital organs from skin surface was approximately 10 mm whilst median distances were of the order of 20 mm. An ultrasonic study was also conducted on 25 subjects on different postures. The postures select were erect, supine and bent forward at the waist at 45°. The effect of breathing cycles on the ultrasound scan was also studied. It was observed that the changing posture and breathing cycle changed position of internal organs up to 80 mm. However, in the case of erect posture with fully breathing cycle, the lower edges of organs such as kidneys and liver were typically 100 mm below the bottom of the rib. It indicated that the body protectors on the torso should extend at the length to protect these organs for serious injuries. Various types of fabric structures were tried [7–12] to make fabric stab, cut and impact resistant. Making fabric stab, cut or impact resistant, role of selecting suitable fibre was paramount [13]. In 1960s, high-strength and high-modulus fibres such as aramid, ultra-high molecular weight polyethylene and Polybenzoxazole (PBO) were developed. When these fibres were used in the fabric, fabric became very strong, with increased ballistic penetration resistance while still maintaining their flexibility. Fabric weave, structural parameters, yarn crimps, etc., also played an important role for getting high-performance fabric out of these fibres. Besides, Shear Thickening Fluid (STF) was also tried [14, 15] to make stab-resistant fabric.

Although numbers of studies were carried out to improve the quality of current body protectors to meet various conventional requirements, still there is a need to further improve its quality considering some newly added threats during riots. It is also required to make it lighter in weight without affecting its quality as it is meant for, so the wearer can walk or run easily. In this study to understand the actual requirement of wearers, a survey is carried out using well-designed questionnaire. In second study, present body protectors were procured from users and industry and tested for stab and impact properties. Finally, to improve upon the stab and impact resistance properties of body protector, various polymeric materials are tried.

2 Materials and Methods

The study was divided into three parts. In the first part, a survey was conducted amongst the paramilitary personnel to understand their difficulties, requirements, etc., related to body protector. In the second study, various samples of body protectors were procured and analyzed using prevailing paramilitary specification as well as VPAM KDIW 2004, Edition 2011 standard. In the last study, some polymeric and composite material were tried to replace the existing material being used in the body protector, to improve upon its stab and impact resistance properties.

2.1 Part-1: Survey of Paramilitary Personal

To get first-hand information about the present body protector from paramilitary personnel, a survey is conducted using well-designed questionnaire. During the designing of questionnaire it was taken into consideration that it would cover all the aspects which influence the safety of soldiers. This questionnaire included the following information related to name, age, length of working hours, native place, problem faced with body protector during riot control operation, etc.

These questionnaires were filled on the basis of interview of individual. A total of 95 personnel including male and female personnel were interviewed.

2.2 Part-2: Analysis of Body Protector Samples

In the second study, 25 samples of body protectors coded as S_1 to S_{25} from users as well as manufacturer ends were procured. Thickness of these materials was determined using screw gauge. The thickness varies from 0.3 to 0.45 mm. The material of various parts of the body protectors was identified using Differential Scanning Calorimeters (DSC Q 20, TA Instrument) and Fourier-transform infrared spectroscopy (FTIR Perkin Elmer UATR Two). The stab and impact properties of these samples were tested as per VPAM KDIW 2004, Edition 2011 standard. As per this standard, the drop mass (5 kg including test tool for stab-knife and 2.5 kg including test tool for impact-block) is allowed to fall on sample of body protector from certain height to simulate striking energy of 65 J and 100 J, respectively. To calculate exact falling height for getting required striking energy, the following formula is used

$$h = \frac{E}{m \cdot g}$$

where E is the striking energy, m is the mass of the knife or block with the holder in kg, g is the acceleration due to gravity and h is the falling height.

The resultant falling height for knife with holder will be 1.33 m for 65 J striking energy and for block it will be 2.04 m for 100 J striking energy.

The penetration depth of the blade in the plasticine is measured with a ruler. Therefore, the resultant hole is excavated. A line diagram of test setup is shown in Fig. 1 for testing stab and impact property of body protector. Figure-2 indicates penetration or deformation of plasticine clay after test. Tests are carried out at 90° angle on the body protector for stab and impact resistance. The striking energy was 60 J and 100 J for Stab (Knife) and Impact (Block) tests, respectively (Fig. 2).

The requirement for pass/fail as per standard is given in Table 1.

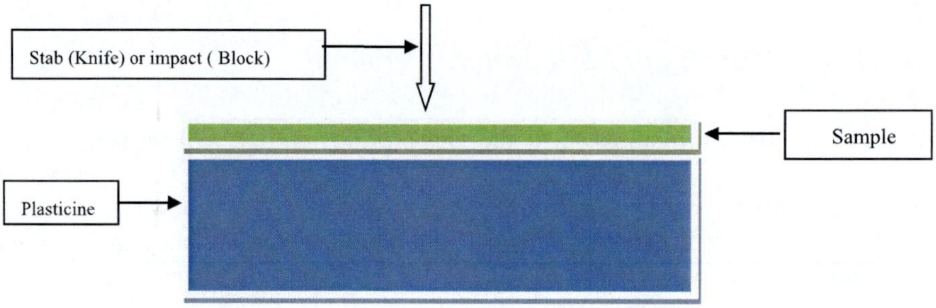

Fig. 1 Testing arrangement stab and impact resistance as per VPAM KDIW 2004 (Edition18.05.2011)

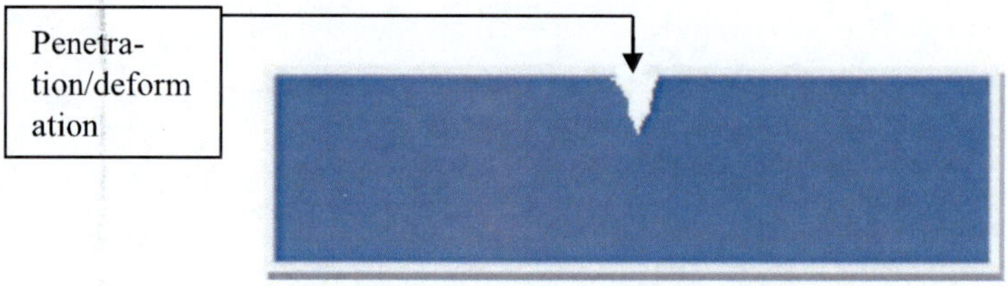

Fig. 2 After penetration of knife/deformation due to impact on Plasticine

Table 1 Stab (Knife) and Impact test requirement

Sr no	Test parameters	Specification	Angle	Type of test	Strike Energy (Joules)	Permitted value (mm)
	Stab resistance test (Knife)	VPAM KDIW 2004:18/05/2011	90^0	Stab-Knife	65	<20 (Penetration depth)
	Impact resistance test (Block)			Impact	100	<20 (Deformation depth)

2.3 Part-3: Improvement of Existing Material to Meet Stab and Impact Resistance Requirement

Three types of materials were tried to improve existing body protector to improve stab and impact resistance property. These materials were coded as B1, B2 and B3. Indian Patent has been filed for these materials (Application No. 201911030134 dt. 25.07.2019). Besides stab and impact tests, they were also analyzed for mass and thickness properties. The mass in g/m^2 of B1, B2 and B3 materials was found to be 350, 410 and 450, respectively, when determined as per IS 7016 part-1 test method. The thickness of these materials was determined using Screw gauge. The thickness of B1, B2 and B3 materials was found to be 0.38 mm, 041 mm and 0.45 mm, respectively.

3 Results and Discussion

3.1 Information Collected by Survey

During the survey one to one interaction was done with 95 respondents using questionnaires. Filled data was analyzed. It was found that.

- 35 respondents of age group 19–30,
- 48 respondents of age group 31–40 and
- 12 respondents were age group of 41–50.

This data indicated that most of the respondents (83 numbers) age was in between 19 and 40 years. It was also revealed that 86% of the respondents were males.

It is clear from the survey (Table 2) that all the respondents were well aware about preventive measures to be taken during a riot/unrest situation. During riots/unrest situation, all respondents informed that they used available protective equipment like body protector, helmet, etc. They also informed that they did not use these equipments regularly. Most of them said that they used these equipments when felt necessary during riots/unrest situation.

The performance of present body protectors is given in Table 3. From the table, it is clear that most of the respondents were not satisfied with the performance of the present body protectors. According to them, these body protectors were heavy and bulky. Besides, these body protectors were also provided with inferior quality of elastic tape and other accessories. In the hygiene point of view, these body protectors smelt bad and were never washed. The design of the present body protector was as per the requirements of the male respondents (troops). The design of present body protectors was not suitable for female troopers. However, both male and female soldiers were not satisfied with the present design and dimensions of the body protectors.

The output of the survey with suggestions for present body protectors is summarized [16] below.

Table 2 Safety practices among Respondents

Parameters	Numbers
Awareness regarding preventive measure	
Yes	95
No	0
Use of available protective equipment	
Yes	95
No	0
Regularity in using any protective equipment	
Regularly	0
When felt necessary	93
Occasionally	0
When it is remembered	2
Not using any protective equipment	0
Safety gadgets used by respondents	
Boots	95
Body protector	95
Helmet with chin guard	95
Protective shield	30

i. The comfort property in the present body protector is not taken into account by the manufacturers/suppliers to facilitate easy movement of the wearer during unrest. The trauma insert made out of polymeric sheets and rubber inserts should not wrap around the whole torso area, especially the area which moves with the body movement. The wearer of body protectors feels uncomfortable when he or she sits. As in the sitting position the lower edges often press firmly against the stomach, hip and side areas of the wearer and makes wearer uncomfortable. Even the top of the chest protector (hard shell) touches the wearer's throat and chin areas and thus making it very uncomfortable. These shortcomings in designing restrict wearer movements while performing any rigorous activities. Therefore, while designing body protector, free body movement of wearers should be considered.

ii. As polymeric sheets and rubber inserts are used in trauma insert, it is not breathable which makes body protector uncomfortable to wear for longer hours especially during summer. One of the ways making it breathable is to create small holes in the polymeric sheets and rubber inserts.

iii. The rubber/soft plastic inserts are being used to protect various body parts such as neck, armpit and lower abdomen areas in the present body protectors, hence providing no protection to the wearer against stab (puncture) attack by knife, screwdriver or spikes. There is a need to use better material which protects the wearers.

Table 3 Performance of present body protectors

Parameters	Numbers
Are you satisfied with body protector?	
Yes	4
No	91
What quality parameters of body protector need to be improved?	
Weight	
• Heavy	81
• Satisfactory	4
• No comment	10
Impact resistance	
• Provide protection	50
• Not satisfactory	45
Elastic tape	
• Working satisfactorily	0
• Become loose after uses	95
Hook & Loop fastener	
• Working satisfactorily	0
• Become unusable after uses	95
Hygiene	
• Bad smell	90
• No problem	5
Washing/cleaning	
• Have you ever washed/cleaned it	No (all 95)
Do you have individual body protector?	No
Ergonomic	
• Comfortable	10
• Not comfortable	85
Do Bulkiness	
• Bulky	90
• No problem	5
Satisfaction with design & dimensions	
• Male	No (all male)
• Female	No (all female)

iv. The foam used in the body protectors is not flame retardant. Due to this, there are chances of catching fire during riot conditions. Nowadays inherent flame retardant foams are available. These can be tried.
v. The weight of body protectors cause significant fatigue to the wearer during maintenance of law and order. It is suggested to use some better and lightweight material to manufacture body protector.
vi. The currently available body protectors are bulky and not designed for female police and paramilitary soldiers. There is a need to have separate design for female wearers.
vii. The woven fabrics which are currently being used to manufacture outer cover of body protectors are neither 'inherent' flame retardant nor cut/slash resistant. Besides, the fabric used for outer cover is not having chemical resistance property. Therefore, there is a need to use such type of outer fabric which is inherently flame retardant, cut/slash resistant and chemical resistant.
viii. The elastic tape and hook & loop fasteners attached with body protectors are of very low quality. These accessories are very important to hold body protector on the body. It was observed that almost all the soldiers were having complaints about its bad quality. They are using a string or rope to hold various parts of body protector. It is required to use better quality of these accessories.

3.2 Analysis of Body Protectors

Body protector or anti-riot suit provides protection against brick batting, cane attack, knife attack, acid attack, Molotov cocktails and projectiles encountered during the riot. The design of body protector shall be such that it should not hinder the movement (walking, running and sitting) of soldiers during operation with minimum fatigue and discomfort. Design shall be ergonomically fit. In this study, 25 body protectors were procured from the paramilitary forces, police forces and manufacturers to analyse various properties including stab and impact resistant.

Various parts of body protector: It was found that body protectors are made using various parts. Trauma pads are kept in pouch made of polyester cotton blend or 100% cotton knitted fabric. Various parts (Fig. 3) and the polymer material used to manufacture these parts (Hard shell and Trauma pad) are given in Table 4.

Stab and impact test: Out of 25 body protector samples, only two samples S_5 (Penetration depth 18 mm) and S_{15} (Penetration depth 19.5 mm) qualified the requirement of Stab (Knife) test. Rest samples have shown penetration of 25 mm–55 mm. The reasons of passing S_5 and S_{15} samples are (i) hard shell of both the samples is made out of Polycarbonate (PC) and (ii) thickness of hard shell of these samples are 0.43 and 0.45 mm, respectively, which are higher than others. It is well-known fact that PC is very tough, impact and stab resistant [17] in nature. This study indicated that there is a need to improve upon the plastic material used in the manufacture of body protectors.

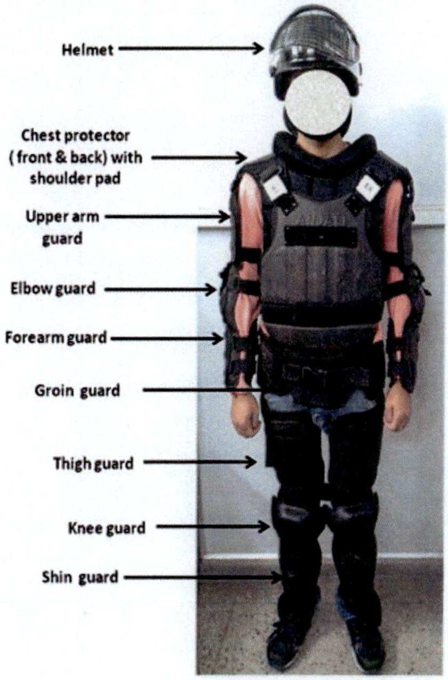

Fig. 3 Parts of body protector

In the case of impact resistance test (using block), 12 samples out of 25 qualified the requirement of deformation depth. The probable reason for passing these 12 samples that they all are made out of combination of Ethylene–Vinyl Acetate (EVA) foam and Polyurethane (PU) foam. EVA foam is described as a specific type of cross-linked closed-cell polyethylene foam. They are designed to be soft, with a rubber-like texture and with good shape recovery after deformation [18]. However, other samples are either having only PU or EPE foams. This test imitates the impact of stone on the body protector during riot. If the deformation depth of plasticine clay is more than 20 mm, it indicates there is a chance of blunt trauma to the wearer of body protector. Blunt trauma can be reduced using a better type of foam, which act as a shock absorber.

3.3 Improvement of Existing Material

The main focus of this study was to improve plastic material used to protect stab and impact resistance property of body protector. For this purpose, experiments were carried out on three types of materials coded as B1, B2 and B3. All the three materials were tested for stab and impact tests. Results of stab and impact tests are given in Table 5. From the table, it is clear that when B1 material was used in the body protector, the penetration depth of knife is 10 mm only. It shows that B1 sample

Table 4 Material used in the present body protectors

Name of the part	Material used in manufacturing Body Protector	
	Hard shell	Trauma pad
Chest and back protector with shoulder pad	Acrylonitrile Butadiene Styrene (ABS), polycarbonates, polyamides (nylons), Polypropylene, Polyethylene	Ethylene-Vinyl Acetate, (EVA), Expanded Polyethylene (EPE), Polyurethane (PU)
Shin guard, upper arm guard, Groin guard, Forearm & elbow guard, Thigh guard with padding material	Polyvinyl chloride/Polypropylene / Polyethylene	Ethylene-Vinyl Acetate, (EVA), Expanded Polyethylene (EPE), Polyurethane (PU)

Table 5 Results of Stab and Impact test of B1, B2 and B3 samples

Sr no	Test parameters	Specification	Test results (mm)		
			B1	B2	B3
	Stab resistance test (Knife) at 90° angle with strike energy 65 J	VPAM KDIW 2004:18/05/201			
	– Penetration depth	1	10	0.0	0.0
	Impact resistance test (Block) at 90° angle with strike energy 100 J		0.0	0.0	0.0
	– Deformation depth				

meets the requirement (<20 mm) of stab resistance test. There was no deformation depth, when the body protector having this material tested for Impact resistance test. It means this sample also qualified the requirement (<20 mm) of Impact resistance test. On the other hand, in the case of B2 and B3 materials, there was no penetration and deformation was found in stab resistance and impact resistance tests, respectively. Hence, these samples also qualified the requirements of stab and impact resistance tests.

4 Conclusions

There is a need to redesign body protector so the wearer can comfortably sit or move to tackle the unruly crowds during riots. The polymeric sheet/rubber insert used in the body protector to resist any trauma during riot should be breathable. The fatigue of the wearer can be reduced by reducing weight of the overall body protector. In the present body protector, rubber/soft plastic inserts are being used to make protective parts of neck, armpit and lower abdomen areas. These rubber/soft plastic inserts are not resistant against stabbing (puncture) attack by knife, screwdriver or spikes. There is a need to use better material in these parts so the wearer can be protected. Besides, there is a need to design body protector for female troopers also.

The elastic tape and hook & loop fasteners are very important accessories to hold the body protector around the body. It is required to improve quality of these accessories.

Out of 25 numbers of body protectors procured from various sources, only 02 body protectors qualified the stab resistance test and 12 body protectors passed impact resistance test. Three materials (coded as A, B and C) were developed under this study. These materials have qualified the requirements of stab and impact tests.

Acknowledgements The authors are thankful to DIG (Prov & Accts) of Rapid Action Force for permitting to carry out the survey amongst their soldiers.

References

1. Horsfall SM, Pollitt J, Belk A, Angood C (1995) ESIS 19. In:. Williams JG, Pavan A (eds) Mechanical Engineering Publications, London, pp 433–442
2. Rouse DA (1994) Patterns of stab wounds: a six year study. Med Sci Law 34(1):67–71
3. Murray LA, Green MA (1987) Hilts and knives a survey of ten years of fatal stabbings. Med Sci Law 27(3):182–184
4. Fligelstone LJ, Johnson RC, Wheeler MH, Salaman JR (1995) An audit of stab wounds in Cardiff. J R Coll Surg Edinb 40:147–170
5. Sparks E (ed) (2012) Advances in military textiles and personal equipment, woodhead publishing series in textiles
6. Horsfall I (2020) Stab resistant Body Armour. https://core.ac.uk/download/pdf/370346.pdf. Accessed 10 May 2020
7. Alpyildiz T, Rochery M, Kurbak A, Flambard X (2011) Stab and cut resistance of knitted structures: a comparative study. Text Res J 81:205–214
8. Miao X, Jiang G, Kong X, Zhao S (2014) Experimental investigation on the stab resistance of warp knitted fabrics. Fibres Text Eastern Europe 5:65–70
9. El Messiry M (2014) Investigation of puncture behaviour of flexible silk fabric composites for soft body armour. Fibres Text Eastern Europe 5:71–76
10. Termonia Y (2004) Impact resistance of woven fabrics. Text Res J 74(8):723–729
11. Song M, Zhang C, Song YuZ, Yan N, Wu Zhu S (2012) Study on the structure and properties of novel impact resistance fabric composites. Appl Mech Mater 182–183:153–157
12. Ertekin G, Marmarali A (2017) Impact resistance behaviour of silicone coated warp knitted spacer fabrics used for protective clothing. J Text Inst 108(12):2123–2131
13. Dimeski D, Srebrenkoska V, Mirceska N (2015) Ballistic impact resistance mechanism of woven fabrics and their composites. Int J Eng Res Technol (IJERT) 4(12)
14. Decker MJ, Halbach CJ, Nam CH, Wetzel ED (2007) Stab resistance of shear thickening fluid (STF)- treated fabrics. Compos Sci Technol 67(3):565–578
15. Gong X, Xu Y, Zhu W, Xuan S, Jiang W et al (2014) Study of the knife stab and puncture-resistant performance for shear thickening fluid enhanced fabric. J Compos Mater 48:641–657
16. Parmar MS, Neha K (2017) Important requirements for anti riot body protector for police and paramilitary forces. Textile value chain, vol 5, p 48
17. Gua Y, Yuan M, Qian X, Wei Y, Liu Y (2020) Rapid prediction of polymer stab resistance performance. Mater Des 192
18. Venkatraman DP, Tyler JD (2020) A critical review of impact resistant materials used in sportswear clothing. https://www.researchgate.net/publication/275648927. Accessed 12 May 2020

Studies on Designing Adaptive Sportswear for Differently Abled Wheelchair Tennis Players of India

N. Bairagi and S. K. Bhuyan

1 Introduction

In India, 21 million people are suffering from disability of one sort or the other according to the census data 2001–2011. Globally, efforts and initiatives are being taken to empower and promote differently able people in all fields. To bring about the mental and physical wellbeing of the differently abled individual, introduction to sports is essential. Among other sports, wheelchair tennis is gaining momentum in urban India where they are trained according to the International Tennis Federation (ITF) rules (IWTT 2016). Wheelchair tennis helps in rehabilitation and also played as a competitive sport [1]. As classified in Paralympic sports the open class category includes players who have paralysis in lower limbs, whereas a quad class can have both players whose both lower and upper limbs are non- functional. The later may not have the ability to manoeuver the wheelchair (ITF) in this case. It was observed that no Indian brands cater to sportswear specific to wheelchair tennis players and the players do have issues with the presently available sportswear. The Indian players presently wear a basic knitted T-shirt and knitted loose/fitted Track pants made out of cotton and polyester, as observed during the pilot study. Therefore, there is a need for functionally designed adaptive sportswear in this category.

The definition of adaptive wear comes from the clothing specifically designed for differently abled people suffering from both physical and mental disorders. Such people can very well adapt to the environment with the help of functional clothing [2]. According to Kabel et al. (2017) clothes can be a factor, which hinders the health

N. Bairagi
Department of Knitwear Design, National Institute of Fashion Technology, 27th Main, HSR Layout, Bengaluru 560102, India

S. K. Bhuyan (✉)
Department of Fashion Technology, National Institute of Fashion Technology, 27th Main, HSR Layout, Bengaluru 560102, India
e-mail: subhalakshmi.bhuyan@nift.ac.in

A. Majumdar et al. (eds.), *Functional Textiles and Clothing 2020*,
https://doi.org/10.1007/978-981-15-9376-5_7

promotion activities to a great extent [3]. Hence, developing adaptive sportswear for wheelchair tennis players by using a user-centric design approach is important in the Indian scenario as there are limited studies in this area. User-Centric Design (UCD) is a process by which the researchers carry out design process by focusing on the users in real-life situations.

A pilot study was done with the wheelchair tennis players at Karnataka State Lawn Tennis Association, Bengaluru, during May 2019. During the study, it was observed that three out of five players have lost the mobility of their legs due to poliomyelitis. According to WHO, Poliomyelitis or Polio is an infectious viral disease that destroys the nerve cells present in the spinal cord causing paralysis in the body. It was found that until 1990, an average of 500 to 1000 children got paralysed daily [4]. This motivated the researchers to take up the study. Therefore, the main aim of this proposed research is to develop design concepts and criteria for adaptive sportswear for wheelchair-bound tennis players of India based on user-centric research approach. The specific objectives of this paper are as follows.

1.1 Objectives

- To study the user needs and requirements in sportswear and develop design criteria and concepts of adaptive functional sportswear for wheelchair-bound tennis players in India who have lost the mobility of their legs due to poliomyelitis.
- To study the thermal imaging of the skin temperature of the players in a real set up using infrared thermography for understanding the thermophysiological comfort requirements with respect to the sportswear of the players.

2 Research Methodology

User-centric design development principles with the design process of Lamb and Kallal (1992) has been used as a theoretical framework for designing of functional adaptive sportswear for wheelchair tennis players, as there are limited researches in this area in the Indian context [5].

Human-centred design research methods or user-centred design methods as classified by Hanington (2003) in three categories: traditional, adapted and innovative methods. First, traditional methods include market research, focus groups, surveys and interviews. The data acquired through these methods provide a good overall view of the design field. Therefore, this method has been adopted and studies on the clothing needs of wheelchair-bound tennis players, the problems faced with the commercially available sportswear were done through interviews with players, coaches, doctors, physiotherapists and caregivers. The second method as mentioned by Hanington 2003 is applied methods, which refers to using research and development methods from design research. Examples of applied methods are qualitative

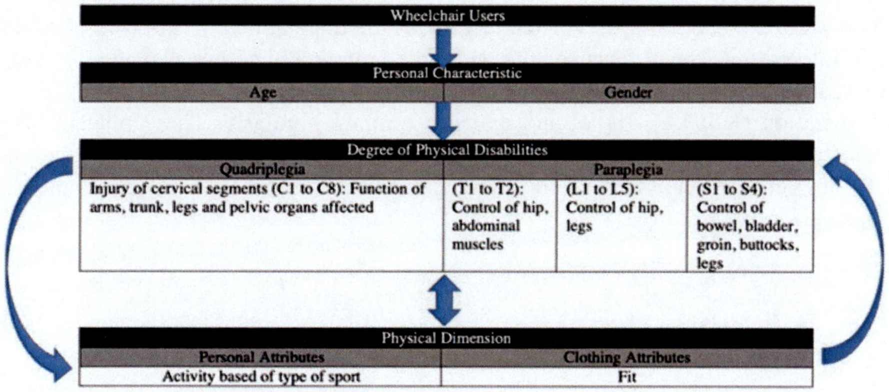

Fig. 1 The proposed model for wheelchair users' sportswear comfort, Lee and Jin (2019)

methods of ethnography, including observations, interviews and interaction methods. Studies in real situation provide qualitative data at how design plays out in practice [6]. These qualitative methods of observing the players playing on their wheelchairs in the tennis field, mapping of changes in the body temperature using thermal cameras in the real set up, formed the base of this study from May 2019—January 2020 at Karnataka State Lawn Tennis Association (KSLTA), Bengaluru, India.

Lamb and Kallal developed a design process that considered the functional needs, expressive needs, and aesthetic needs of the clothing consumer. Functional needs include fit, mobility, comfort, protection, and donning/doffing of the garment. Expressive needs include values, roles, status and self-esteem of the consumer. Aesthetic needs include art elements, design principles, and body/garment relationships [5]. As these needs are also important and applicable for designing sportswear for the wheelchair-bound tennis players, it has been studied through the interviews and observations. Additionally, the conceptual design framework for wheelchair users' sportswear comfort as proposed by Lee and Jin (2019) has also been used for conceptual design development, as shown in Fig. 1. The conceptual design framework for wheelchair users' sportswear comfort based on Lee and Jin (2019) takes into account the degree of physical disability, type of sports and personal attributes into consideration to cover functional, physiological and psychological comfort [7].

2.1 Participants

The study was conducted with one male and one female wheelchair tennis players suffering from poliomyelitis, who volunteered for the study. Male and female players were selected as literature reports that female wheelchair players have a more positive attitude towards adaptive sportswear than male wheelchair-bound players [8]. The experience of playing wheelchair tennis by the male athlete is 3 years and the female athlete is 8 months. Till January 2020, the male athlete has played National and International level tournaments while the female athlete has played state-level tournaments. Both the athletes are associated with Karnataka State Lawn Tennis Association (KSLTA), India, and undergo training regularly. These players have been trained by national and international coaches in Bengaluru. Every weekend the athletes get together at KSLTA for wheelchair tennis matches. Informed consent was taken prior to the study.

2.2 Data Collection

The qualitative and quantitative data have been collected initially through observations, the structured questionnaire with users and stakeholders (coaches, doctors, physiotherapists, sportswear experts, etc.) using Lamb and Kallal's (1992) functional, expressive and aesthetic (FEA) consumer needs model. During the interview, the researcher tried to find out the problems the players faced while playing wheelchair tennis especially with players affected by poliomyelitis. The interview questions were divided into the following groups: (a) difficulties with the present sportswear (b) FEA needs and (c) additional information on the needs and requirements. The research approach has been presented in Fig. 2. Visual data in the form of photographs and videos, verbal data and textual data has been collected during the study in repeated phases from May 2019 to January 2020 at Karnataka State Lawn Tennis Association (KSLTA), Bengaluru, India. Narrative recording and field notes were taken while conducting the observation technique.

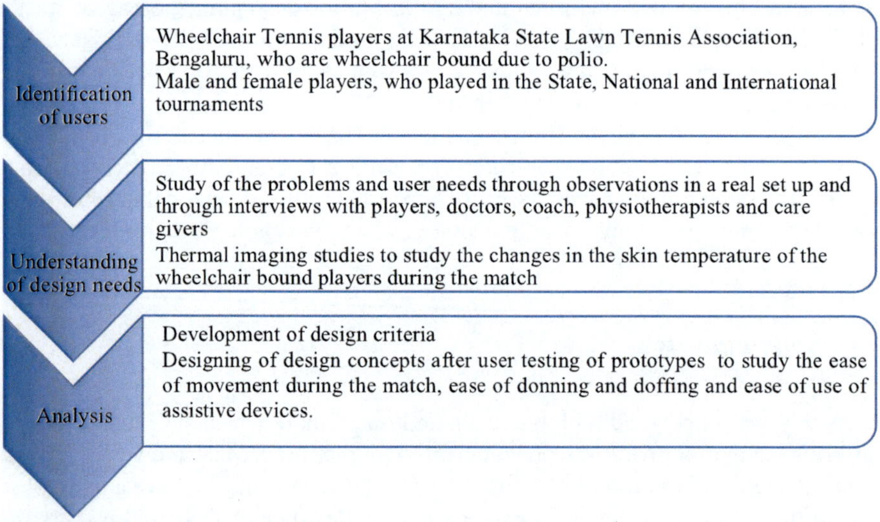

Fig. 2 Research approach

The adaptive sportswear prototypes were assessed for the movements of the wheelchair during the game; interference with the assistive devices used by the wheelchair-bound players, ease of donning and doffing and on the muscle strain faced by the wheelchair-bound players during January 2020. Data was collected through verbal feedback and video recordings for analysis. The modifications were made on the prototypes based on the analysis, prior to proposing of design concepts.

2.3 Infrared Thermography Studies

Additionally, infrared thermography was conducted on the female player in a real set up at Karnataka State Lawn Tennis Association, Bengaluru during January 2020. Tests were carried out in an open court at Karnataka State Lawn Tennis Association with a recorded temperature of 28 °C and average relative humidity of 76 %. The aim of the thermal imaging is to study the changes in the skin temperature of the wheelchair-bound players during the match using Fluke Ti480 Pro Thermal Imaging Camera with the resolution of pixels 640 × 480 and with the sensitivity of 0.05 °C at 30 °C target temperature. Thermal images of the sportsman body front and back were taken after 6, 12 and 18 min of the match to observe the process of the rise of cutaneous temperature and the thermoregulation with respect to different zones of the body. The results were analysed using Fluke Connect software, focusing on the muscle zones according to the model proposed by Fernandez-Cuevas et al. [9]. The results were analysed for the development of design criteria of adaptive sportswear.

3 Research Findings

The observations from the user-centred studies, findings from the in-depth interviews with the sportsman has been analysed and presented in this section. Additionally, the data from the infrared thermography studies representing the changes in the skin temperature of the wheelchair-bound players during the match were analysed for the development of design criteria of adaptive sportswear. Findings from the research observations have been categorised into different areas of functional needs based on: the movements of the wheelchair during the game, on the assistive devices used by the wheelchair-bound players who have lost the mobility of their legs due to poliomyelitis, on the muscle strain faced by the wheelchair-bound players, on studies of infrared thermography of the wheelchair-bound players and on common strain and injuries encountered by the wheelchair-bound players. All needs have been discussed in the following sections. The design criteria developed (Table 1) has been used in conceptualisation of the designs of the adaptive sportswear for wheelchair-bound tennis players presented in Sect. 3.7.

Table 1 Design criteria for adaptive sportswear for wheelchair-bound tennis players

Clothing	Characteristics of the adapted sportswear
General characteristics	· Specific opening for the player to easily put on and take off · Need to have easy fasteners (zips, elastic bands, easy to use buttons, string cords) · Avoid friction between sportswear and chair · Allow independent dressing and undressing · Allow independent use of the washroom · Aesthetically appealing for psychological comfort and motivation to play · Concealed design modifications, that it would not be visible to others
Upper body garment	· Allow maximum movement of the trunk and upper limbs · Reduce drag and resistance to upper arm movement · Facilitate the process of thermoregulation to avoid fatigue, through quick wicking of perspiration in the different zones like shoulders areas, upper arms, below bust area, sides of the trunk, thighs and crotch areas · Have longer length at the back to prevent the rise of sportswear during the match
Lower body garment	· Need to have a higher back rise than usual track pants · Wider waist band to prevent slippage during the match · Should be adjustable to the leg length · Should facilitate the use of braces and calipers · The lower should provide enough stretch and room at the buttock and suitable fabric to prevent excessive heat buildup and sweating through quick wicking · Should be free from thick and hard seams to prevent pressure sores at the back and buttock areas. · Abrasion-resistant fabric
Materials	· Be absorbent with quick-wicking properties, enhanced moisture vapour permeability and air permeability for moisture management · Easy to maintenance · Good abrasion resistance · Minimum body odour retention

3.1 Functional Needs Based on Movements of the Wheelchair

During the pilot study, it was observed that in wheelchair tennis, one arm is used to hold the racquet and the other arm is utilised to manoeuver the sports wheelchair (Fig. 3). The tennis racquet is an additional constraint to the player during propulsion in wheelchair tennis as observed and as also reported in the literature [10]. Basic strokes in this game that contributes to their performances are serve, forehand, backhand, forehand volley, backhand volley and overhead smash.

Fig. 3 Sketch of a female wheelchair-bound tennis player

Wheelchair tennis players can have specific movements such as start/sprinting/brake and turning (pivoting). Wheelchair is also treated as an integral part of player's body and their buttock should always be in contact with the seat of the wheelchair during the play as per ITF rules. Therefore, the adaptive sportswear upper body garment and lower body garment need to facilitate these ergonomic needs to ease the movements while using the wheelchair.

3.2 Functional Needs Based on Assistive Devices Used by the Wheelchair-Bound Players

During the study, it was observed that one of the main needs in the sportswear for wheelchair tennis players who have lost the mobility of their legs due to poliomyelitis are functional needs to resolve or eliminate barriers during the sports, such as minimum interference with the calipers or braces used by the players on their legs. Additionally, easy donning and doffing of the sportswear to access washrooms. The Indian players are dependent on the assistants due to the absence of adaptive sportswear. Therefore, the lower body garment of the sportswear needs to be designed to accommodate calipers and braces as the present lower body garments/track pants do not consider the calipers for the polio bound players.

Wheelchair-bound tennis players require a high level of aerobic fitness as the movements can be both intermittent and multidirectional [1]. The velocity of the wheelchair of professional players ranges from 4.3 m/s, which is reduced by 0.18 m/s during the racquet condition [10]. Literature reports an average heart rate of a wheelchair-bound tennis player is around 114 beats/min [11]. Additionally, wheelchair players need an abdominal binder to minimise venous blood pooling. Pressure garments such as stockings have been recommended by researchers which also reduces the risk of venous blood pooling [7]. The participants of this study also used an abdominal strap, strap over thighs and ankle straps to secure their foot on the wheelchair, so that it does not touch the ground while playing. Therefore, the sportswear design needs to take this into consideration to prevent added discomfort and facilitate both intermittent and multidirectional movement with the assistive devices.

3.3 Functional Needs Based on Muscle Strain of the Wheelchair-Bound Players

Structured interviews with the stakeholders (Doctor, Sports Physiotherapist and Coach at Karnataka State Lawn Tennis Association) helped to gain insights into the physiological aspects of the players. While playing in a wheelchair, the hip is stable, but the shoulder is flexible and in movement. According to Fabrocini (1995), the performance of a player depends on the trunk rotation that is quite significant for these players who even need to balance themselves to control the shots. The trunk muscles help in the kinematic chain to move the body [12]. Ample amount of strain/stress occurs in the upper body and lower back [13]. It has been observed that the strain on the upper torso is more in case of wheelchair-bound players as it endures more pressure to keep the balance and accomplish the act of playing. Therefore, while designing adaptive clothing the sportswear upper need to provide the required compression to support the strain on the torso and also facilitate the thermophysiological regulation.

3.4 Functional Needs Based on Studies of Infrared Thermography of Skin Temperature of the Wheelchair-Bound Players

Infrared thermography is a non-invasive and easy-to-perform method of imaging of the body temperature in athletes and sportsman. It helps us to understand the skin temperature rise and thermal adaptations of the sportsman. This, in turn, helps in designing of the sportswear as the temperature rise in specific zones of the body can be visualised during the activity in a real set up and the sportswear can be designed accordingly for better thermoregulation, which can, in turn, improve performance, of the wheelchair-bound players by reducing fatigue. Images have been recorded at 6 min from the start of the match, then at 12 min and then at 18 min of the female wheelchair-bound player. The observations have been presented in Figs. 4, 5 and 6.

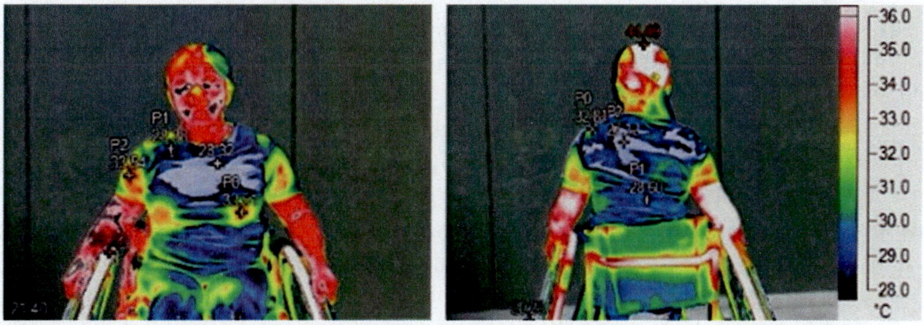

Fig. 4 Thermal imaging of the front and back of the wheelchair-bound female player after 6 min from the start of the tennis match

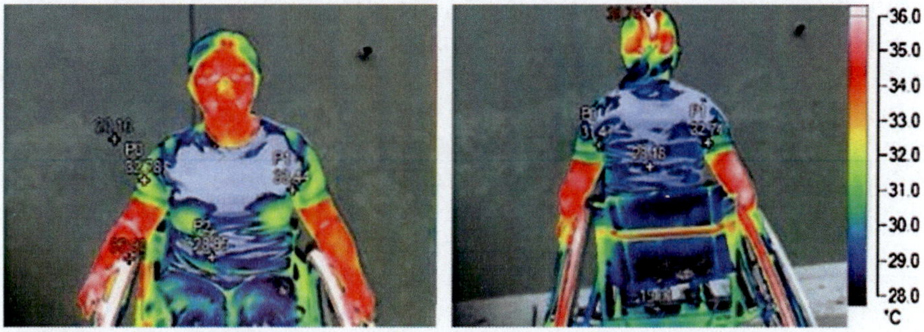

Fig. 5 Thermal imaging of the front and back of the wheelchair-bound female player after 12 min from the start of the tennis match

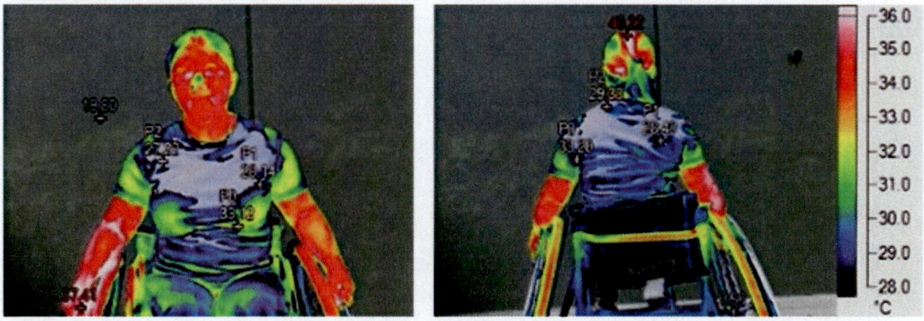

Fig. 6 Thermal imaging of the front and back of the wheelchair-bound female player after 12 min from the start of the tennis match

At the starting of the match, after 6 min, it can be observed from Fig. 4, that the hot spots in the player's body, which records a temperature above 33 °C, are located in the shoulders areas, upper arms, below bust area, sides of the trunk, thighs and crotch areas. This may be attributed to the muscles which are more functional during the activity of playing wheelchair tennis. The main muscles involved are trapezius in the shoulders and back, deltoid in the upper arms, biceps, serratus anterior in the lower bust, Latissimus in the back and Sartorius, Rectus femoris, Gracilis and Adductor longus are the main muscles in the thigh area. Therefore, these zones recorded a higher temperature in the player due to an increase in cutaneous blood flow. The minimum temperature is 28 °C. This is also due to the use of wheelchair, straps and calipers that restricts the airflow in a sitting position. Properly designed adaptive clothing, taking the hot spots of the body into consideration, can reduce the rise in the cutaneous skin temperature and reduce fatigue of the sportswear.

In Fig. 5, where the infrared thermal imaging is taken after 12 min of the match, it could be observed that the zones which recorded a previous temperature of above 33 °C, drops in the range of 31–32 °C due to the process of thermoregulation that sets in the players. In Fig. 6, it can be observed that the temperature of the hot spots further drops to 30–31 °C as the body tries to achieve thermoregulation. The thermal imaging study of the player also showed the effect of physical exercise on the rise in the skin temperature during the match. The rise in the skin temperature is due to the increase in the cutaneous blood flow (CBF) during the exercise and is linked to an individual level of vasodilation and vasoconstriction [14]. Johnson 1992 and Robinson 1963 reported that at the beginning of the exercise the demand of blood flow to the working muscles cause vasoconstriction, but as the body temperature rises, the thermal regulatory processes predominate and the skin blood vessels dilate, perspiration occurs, increasing heat dissipation through the skin by irradiation and evaporation [14, 15]. This needs to be catered by the sportswear upper body garment and lower body garment through designing with open structured fabric at those specific zones that would enhance quick-wicking, moisture vapour permeability and air permeability. Therefore, the designing of the sportswear need to use fabrics which facilitates quick-wicking, moisture vapour permeability and air permeability with the required level of the stretch as presented in the design criteria (Table 1).

3.5 *Functional Needs Based on Common Strain and Injuries of the Wheelchair-Bound Players*

Paralysis makes the calf and hamstring muscles weak and significant loss in muscle mass occurs in case of poliomyelitis. This impacts getting the right size of lower that is fitted and functional. It was observed that the wheelchair-bound players need a stronger back to support as they play. Regular exercises of upper arm and shoulder strengthen the body and help to prevent injuries. But while trainings, the wheelchair-bound tennis players often experience injuries such as shoulder sprain/blisters due to abrasion/sprains/ muscle spasm/soft tissue injury in shoulder, arm, wrist etc. Similar observations have been reported in the literature [16]. Along with the upper torso, the lower back also undergoes a considerable amount of strain. Many players also develop pressure sores and need appropriate intervention in the sportswear to reduce such problems. The main issues faced by the wheelchair-bound tennis players of India are:

While playing, the skin of the elbow gets abraded against the tires of the wheelchair hence additional protection may be incorporated while designing of the upper.

- It has been observed that the palms of these users become hard due to the act of rotating the push rim to propel the wheelchair.
- The straps present in the wheelchair to secure the player, in abdomen, thigh and ankle areas sometimes create rashes in those parts of the body.
- The wheelchair tennis players find it difficult to don and doff their lowers and use washrooms independently.
- The wheelchair tennis players affected by polio have fit issues with the lowers available in the commercial sportswear. Many of the players have weak muscles in the lower limb and muscular atrophy that needs to be taken care of while designing adaptive sportswear.
- Depending on the excessive sweating and heat generation in certain areas of the body.

3.6 Development of Design Criteria for Adaptive Sportswear for Wheelchair-Bound Tennis Players

The problems and needs stated by the users and stakeholders during the ethnographic study and structured interview helped in initial design conceptualisation. The main problems faced by the players are as follows:

- Tight armhole of the upper creates difficulty in hand movement.
- Lowers need to accommodate calipers, otherwise, removal of calipers while playing makes them unstable.
- While playing, the back part of the T-shirt lifts up exposing the skin that comes in contact with the wheelchair back. That causes abrasion and irritation in the lower back as it is in contact with the wheelchair.
- While playing the waist of the trouser tends to slip down. To prevent such slippage, the player needs the appropriate design of the waist band.

The main aesthetic requirements of the players are as follows:

- Male and female players have individual colour preferences.
- Both male and female players would prefer to have sportswear with concealed modifications that would not be visible to others.
- Female players would prefer sportswear that would hide the physical deformity of the body.

The analysis of the activities of the players, environment, the interaction of the player with the wheelchair and assistive devices were also done prior to design conceptualisation. The findings were further discussed with stakeholders and users before ideation of proposed concepts. The design criteria developed has been presented in Table 1.

3.7 Development of Design Concepts of Adaptive Sportswear for Wheelchair-Bound Tennis Players

Based on the above findings, eight tested design concepts for the sportswear upper body garment and lower body garment are presented in this section. The details of the conceptualised designs have been presented in this section. The designs were user-tested through wearer trial to study the ease of movement during the match, ease of donning and doffing and ease of use of assistive devices.

Upper garment with raglan styled sleeves has been designed for easy movement of the hands of the players. Mesh panels added on to the sleeves and near both side, seams to facilitate thermoregulation. Open knit structures are added near the abdomen where sweating is excessive as observed in the infrared thermal imaging. Back pattern is longer than the front pattern so that it does not expose the skin of the player while playing as presented in Fig. 7.

Option 2 is presented in Fig. 8, presents knee-length tennis dress (whole-body garment) with mesh panels on the sleeves, near the side seams and at the back with the additional panel in the abdomen area to facilitate thermoregulation in female wheelchair-bound tennis players. The zipper has been attached on the left side seam from below the armhole to the bottom hem for easy dressing and undressing.

Racer back T-shirt with in-built bra cups is presented in Fig. 9, It would help to avoid wearing an extra layer of the sports bra. Racer back style also helps in easy arm movement. Laser-cut panels in the shoulder, back and below the bust would

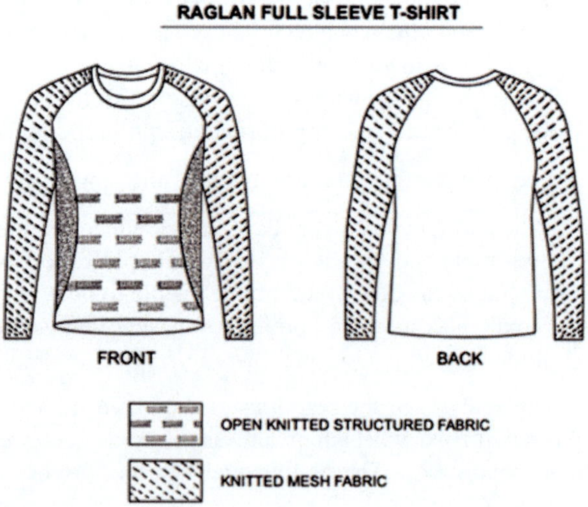

Fig. 7 Design concept 1: Full sleeve raglan T-shirt

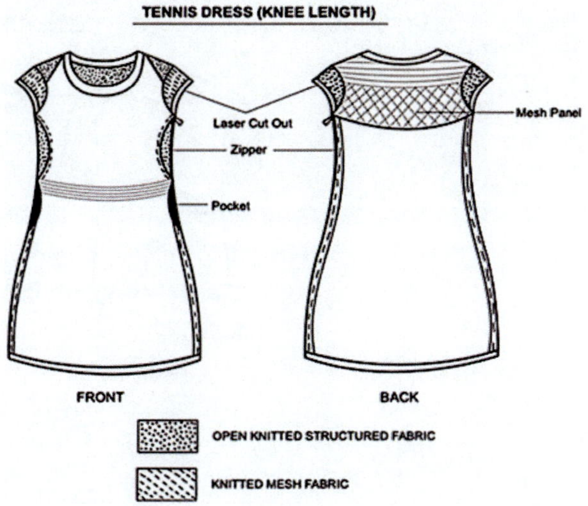

Fig. 8 Design Concept 2: Tennis Dress (whole-body garment)

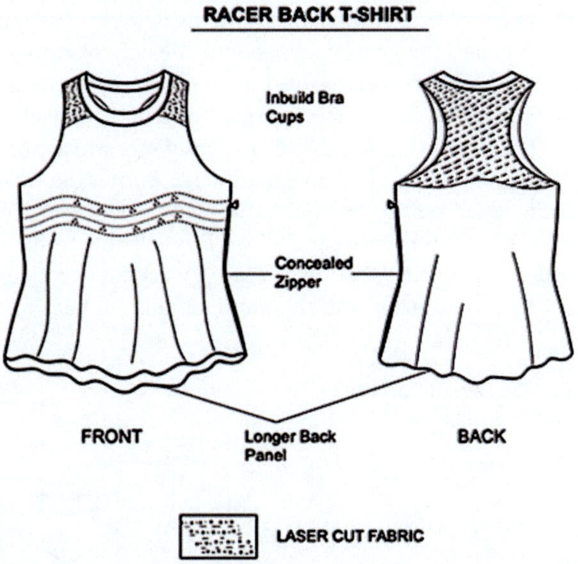

Fig. 9 Design Concept 3: Sleeveless Racer back T-shirt

facilitate better thermoregulation of the hot spots in the body. Zipper added on the left side seam for easy donning and doffing.

Half sleeve raglan T-shirt presented in Fig. 10 has a raglan sleeve for easy movement of the arms. Mesh panels added on the sleeves and below the bust. It also has a laser-cut panel at the back based on the thermal zones in the body of the wheelchair tennis player. The bottom hem has elastic so that it holds on to the waist. It also has straps with buttonholes on both sides of the side seams. It may be attached to the button on a track pant to avoid lifting upward during the match.

Fig. 10 Design Concept 4: Half sleeve raglan T-shirt

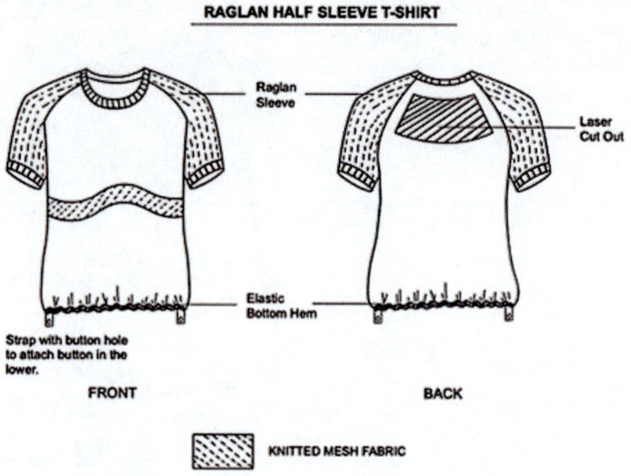

Design concept 5 (Fig. 11) shows track pant (option 1) with broad elastic waistband. The back rise is made to rise higher than a usual track pant keeping in mind the needs of the wheelchair user tennis player. The out seams on both the sides have mesh panels that can be zipped with the help of zipper running from waistband to the bottom hem. It has extra space to accommodate calipers if required. Mesh panels have been added even on the inseam at the front and back to prevent excessive sweating.

Track pant (Option 2, Fig. 12) has wider waistband and has been designed to give extra comfort and to prevent slippage during the match. Open knit structures have been added on the abdomen, inseam and outseam areas for ventilation. Drawcord added inside the waistband which is not visible from outside. Back rise is higher than usual. Track pant (Option 3, Fig. 12) has flat waist band with mesh panels added on the thigh and back near the popliteal and on inseam and outseam areas. Open knit

Fig. 11 Design Concept 5: Track Pant Option 1

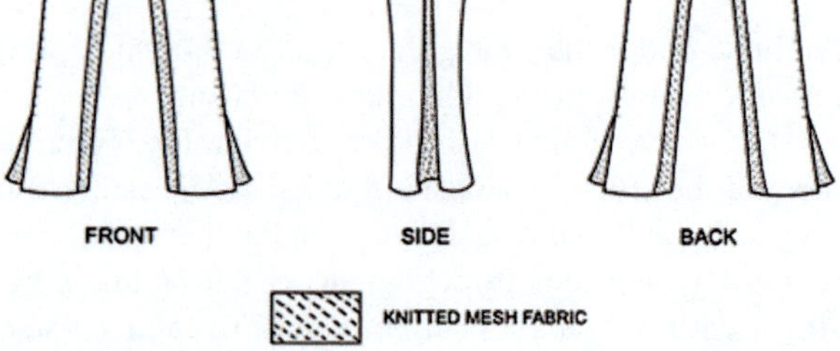

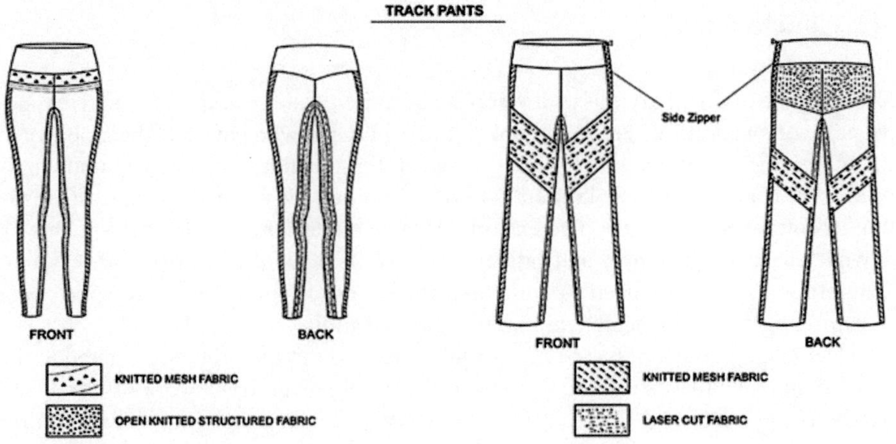

Fig. 12 Design Concept 6 and 7: Track pants Option 2 and 3

structures attached at the buttock area. Zipper attached on the left side of the track pant. High back rise pattern has been incorporated in the track pant.

Design Concept 8 (Fig. 13) presents unisex jogger with mesh panels on inseam and outseam. Elastic added in the waistband and bottom hem. High back rise and concealed zipper on the centre front have been incorporated in the design.

The designs developed need to be custom made according to the size of the players as wheelchair-bound players generally have more developed muscles in the upper torso than in the lower limbs [17]. Additional thermography studies are required on the designs developed in the real set up to study the efficiency of the changes in the skin temperature of the wheelchair-bound players during the match.

Fig. 13 Design Concept 8: Track pant Option 4

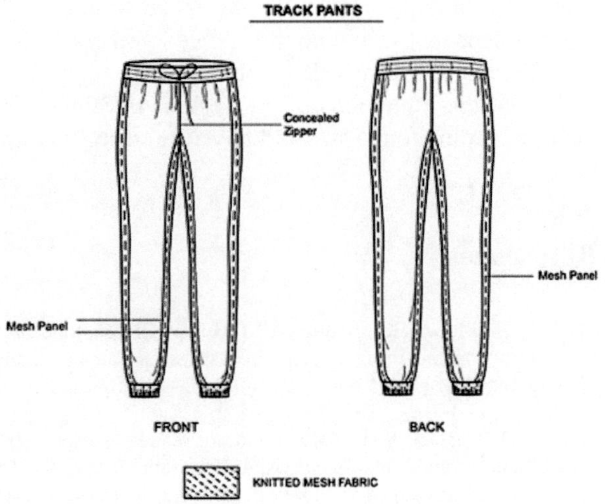

4 Conclusions

The purpose of this study was to develop design criteria and concepts of adaptive functional sportswear for wheelchair tennis players in India, who have lost the mobility of their legs due to poliomyelitis. The sportswear worn by the wheelchair-bound tennis players is generally a basic knitted T-shirt and knitted loose/fitted Track pants made out of cotton and polyester. The existing garments interfere with the movements on a wheelchair and the braces and calipers used by the players, apart from other issues. Therefore, there was a need to undertake studies on designing adaptive sportswear for differently abled wheelchair tennis players of India.

User-centric research has been carried out to find out the problems related to the players in real situations along with the conceptual design framework of Lamb and Kallal (1992). Lee and Jin (2019) model for wheelchair users' sportswear comfort has also been used for design development. Additionally, infrared thermal imaging (thermography) techniques were used to study the changes in the skin temperature of the wheelchair-bound players during the match in a real set up, prior to conceptualisation. The prototypes developed were user-tested prior to proposing of design concepts.

It was observed that during the match the cutaneous temperature of the players the rise up to 33–34 °C in certain zones like shoulders areas, upper arms, below bust area, sides of the trunk, thighs and crotch areas. This may be attributed to the muscles which are more functional during the match and hence have higher cutaneous blood flow. The minimum cutaneous temperature was observed to be 28 °C. The hot spots are also due to the use of wheelchair, straps and calipers that restricts the airflow in a sitting position. This needs to be taken into account while designing of sportswear in addition to functional designs that facilitate the movements of the wheelchair during the game, use of assistive devices used by the wheelchair-bound players, reduce muscle strain and injuries faced by the wheelchair-bound players. The aesthetic needs of both male and female players are similar in the aspect of concealed design interventions that would not be visible to others. A female player on the other hand would prefer sportswear that would hide the physical deformity of the body as well.

One of the limitations of the designs developed is that they need to be custom made according to the size and physical attributes of the wheelchair-bound players.

References

1. Roy JL, Menear KS, Schmid MM, Hunter GR, Malone LA (2006) Physiological responses of skilled players during a competitive wheelchair tennis match. J Strength Cond Res 20(3):665
2. Boorady LM (2011) Functional clothing- principles of fit. Indian J Fibre Text Res 36(4):344–347
3. Kabel D, Black KM (2017) Clothing related barriers experienced by people with mobility disabilities and impairments. Appl Ergonom 59(Part A):165–169
4. John TJ, Vashistha VM (2013) Eradicating poliomyelitis: India's journey from hyperendemic to polio free status. Indian J Med Res 137(5):881–894

5. Lamb JM, Kallal MJ (1992) A conceptual framework for apparel design. Cloth Text Res J 10:42–47
6. Hanington B (2003) Methods in the making: a perspective on the state of human research in design. Des Issues 19(4):9–18
7. Lee H, Jin H (2019) Conceptual design framework as a model for wheelchair users' sportswear comfort. Fashion Text: Int J Interdiscipl Res 6:23. https://doi.org/10.1186/s40691-019-0179-z
8. Wingate SB, Kaiser SB, Freeman CM (1986) Salience of disability cues in functional clothing: a multidimensional approach. Cloth Text Res J 4(2):37–47
9. Fernandez-Cuevas I, Sillero-Quintana M, Garcia-Conception MA, Ribot-Serrano J, Gomez-Carmona P (2014) Monitoring skin thermal response to training with infrared thermography. New Stud Athlet 29(1):57–71
10. Goosey-Tolfrey V, Moss AD (2005) The velocity characteristics of wheelchair tennis players with and without the use of racquets. Adapt Phys Act Q 22:291–301
11. Croft L, Dybrus S, Lenton J, Goosey-Tolfrey V (2010) A comparison of the physiological demands of wheelchair basketball and wheelchair tennis. Int J Sports Physiol Perform 5:301–315
12. Fabrocini B (1995) The planning of a powerful trunk for tennis. Strength Cond:25–29
13. Kibler WB, McQueen C, Uhi T (1988) Fitness evaluations and fitness findings in competitive junior tennis players. Clin Sports Med 7:403–416
14. Johnson JM (1992) Exercise and the cutaneous circulation. Exerc Sport Sci Rev 20:59–97
15. Robinson S (1963) Temperature regulation in exercise. Pediatrics 32(Suppl):691–702
16. Ferrara MS, Davis RW (1990) Injuries to elite wheelchair athletes. Paraplegia 28:335–341
17. Chang WM, Zhao YX, Guo RP, Wang Q, Gu XD (2009) Design and study of clothing structure for people with limb disabilities. J Fiber Bioeng Inf 2(1):62–67

Functional Printing and Finishing

Photoluminescent Printed Fabrics: Design and Development of Home Fashion Products to Aid Nighttime Navigation

R. Sharma, N. Bairagi, and M. Gupta

1 Introduction

Photo-luminescent (PL) pigments are rare earth metals that absorb light and emit it slowly as a pale blue/bluish-green light over 6–8 h after initial excitation from an external light source. They may be incorporated in the textile as filaments, coating, tapes or vinyl stickers [1–5]. In our previous paper, the effect of particle size and concentration of the PL pigments on textiles by print method on the intensity of luminance or emitted light was discussed [6]-. This paper focuses on the effects of PL pigments on various textiles with respect to morphology and its use in various end uses as home fashion products, especially in the area of elderly care.

Light is a powerful modulator of cognition through its long-term effects on circadian rhythm and has direct effects on brain function [7]. Many sleep specialists suggest that widespread exposure to blue light is a major contributor to insomnia in general [8]. In the elderly, insomnia and cognitive impairment are major factors for fall and it has far-reaching consequences for the independence of the elderly. Those affected become increasingly insecure about moving around. Among older adults, the most common reasons [9] for accidentally falling or slipping include wet or recently polished floors, such as in a bathroom, dim light, rugs or carpets that are not properly secured, reaching for storage areas, such as cupboards and stairs. There

R. Sharma (✉)
Department of Textile Design, National Institute of Fashion Technology, HSR Layout, Bengaluru 560102, India
e-mail: richa.sharma@nift.ac.in

N. Bairagi
Department of Knitwear Design, National Institute of Fashion Technology, HSR Layout, Bengaluru 560102, India

M. Gupta
Nanophotonics Lab, Department of Physics, Indian Institute of Technology, New Delhi 110016, India

A. Majumdar et al. (eds.), *Functional Textiles and Clothing 2020*,
https://doi.org/10.1007/978-981-15-9376-5_8

is limited research on developing home fashion products like pillows, runners, rug and carpets to aid nighttime navigation.

Cognitive impairment [10] when eyesight deteriorates due to age or illness, it has far-reaching consequences for the independence of the elderly. Light cannot correct deteriorating eyesight, but it can help to compensate, restoring self-confidence. Proper illumination helps avoid visual missteps and prevent falls. Therefore, designing home furnishings with photoluminescent pigment textile prints may be an area of study that may enable the aged people to function with ease. Additionally, there are limited studies on a real nighttime environment to evaluate the effectiveness of the photoluminescent printed textiles by elderly people. Thus the objectives of the paper may be briefly presented as:

- To study the effect of printing with PL pigments of different pigment concentration and particle size on luminosity.
- Also, to correlate the luminosity of PL printed textiles with the visual perception of the elderly in a nighttime environment.
- To design home fashion products to aid edge definition and path illumination facilitating nighttime navigation.

2 Research Methodology

The research study was conducted using mixed methods. It consisted of experimental research to study the effects of PL pigments printed textiles using lux meter and SEM imaging, followed by a study of visual perception by users in a nighttime environment. This was followed by design development using user-centric research approach.

3 Materials and Methods

The methodology used is experimental research to study the emission intensity of the printed textiles and a visual perception study by elderly people in a nighttime environment in a real-time natural set up. To study the intensity of photoluminescence and effects of different photoluminescent pigments printed textiles lux meter and Time Resolved Photo-luminescent (TR-PL) spectroscopy were used. This was followed by a study of visual perception by the elderly in a nighttime environment in their personal bedrooms.

The research study was carried out on fabrics popularly used for home furnishings in India. Therefore, 100 % cotton, medium weight plain-woven fabric of 300 g/m^2, with EPI 68 and PPI 68 was used. Additionally, studies were also carried out on different types of fabrics commercially for home fashion namely mulberry silk (plain woven, 40 g/m^2), dupion silk (plain woven, 60 g/m^2) and matka silk (plain woven, 100 g/m^2) to study the effect of the luminosity of the photoluminescent pigments.

Table 1 Experimental plan with variable concentration of PL Pigment on cotton fabrics

S. No	Pigment	Pigment Sample code	Particle size (μm)	Sample code for the variable concentration of pigments in the print paste		
				5 % (C3)	10 % (C4)	20 % (C5)
1	Yellow–Green	G1	10–15	G1C3	G1C4	G1C5
2		G4	50–60	G4C3	G4C4	G4C5
3	Aqua Blue	B1	10–15	B1C3	B1C4	B1C5
4		B4	50–60	B4C3	B4C4	B4C5

The two photoluminescent pigments used for the study were composed of SrAl2O4: Eu^{2+}, Dy^{3+} with a yellow-green (G) hue having λmax of 520 nm and another with aqua blue (B) hue with λmax at 490 nm. The particle size of the pigments used was in the range of 10–15 μm and 50–60 μm, sourced from Jash Marketing, India. Cotton fabric samples were initially printed using 3 variable concentrations of pigment (5, 10 and 20 %) based on the weight of the print paste, with two different particle sizes in each aqua blue and yellow-green pigment as presented in Table 1. All the samples were industrially screen-printed (mesh size 200) with a curing temperature of 120 °C. Printed samples were used for study and observations.

Additionally, to study the decay of intensity, Time-Resolved Photoluminescent (TRPL) spectroscopy was carried out with higher precision spectrophotometer QE pro. The excitation wavelength used was 480 nm, LED single point light source of 220 V. The experimental setup consisted of a direct probe, without any filters (so as to capture emission without any loss) to the fiber optic connected to spectrophotometer QE Pro equipped with Igor-Pro, data management software to obtain spatial intensity maps and PL intensity time scans.

The visual perceptions of the printed photoluminescent samples were correlated with the lux readings. For this, a user-centric approach was taken to study the visual perception by users in real-life environment. The user study was conducted with the elderly using non-probability, convenience sampling method, as the studies had to be conducted in the users own home setting. Prior consent was taken from the elderly respondents to be a part of the study. The experiment was conducted with 10 elderly (4 men and 6 women) in the age of 60–75 yrs in their personal bedrooms in the night to ascertain the user experience in the real-time environment.

The photoluminescent pigment printed cotton samples with 4 broad printed stripes of 20 % pigment concentration and different particle sizes, that is, G1C5, G4C5, B1C5 and B4C5 were placed as curtains at the window (see Fig. 1) and users were requested to follow their usual routine in the night. Printed samples of 5 and 10 % concentration were not included in this study as the emission intensity was lower than that of 20 % pigment concentration. All the participants were interviewed the following morning. This was repeated for three consecutive days to obtain consistent data from the users. The users were asked to rate the visual perception of the four printed bands in the curtain on a scale of 1–5 and also to propose suitable applications

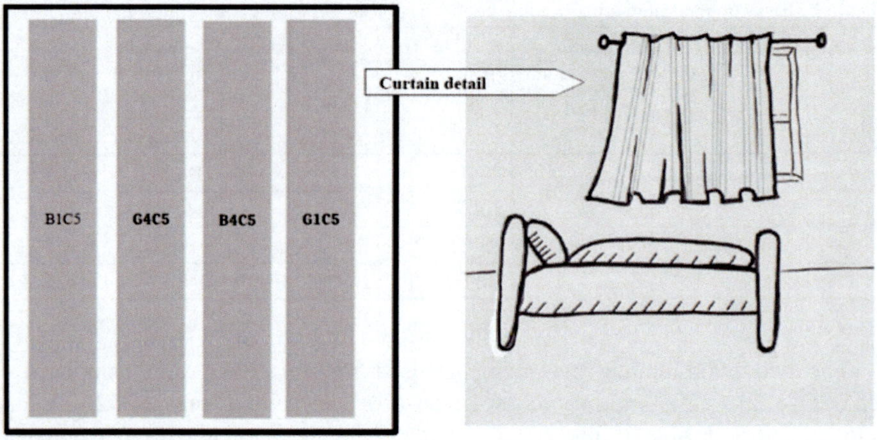

Fig. 1 The PL pigment print curtain detail on the left; and the bedroom visualization for the visual perception assessment

of the printed textiles. The average rating has been considered for the study. User preferences accounted for the final conceptualization of product design and proposed applications for end use.

4 Results and Discussion

4.1 Effect of Particle Size and Concentration of Photoluminescent Pigment Printed Textiles on Luminescence

It was observed that both pigment size and concentration plays a significant role in the intensity of luminescence exhibited by the cotton printed textiles. The absolute lux readings using an illuminometer are exhibited below (see Fig. 2). Aqua blue pigment (B4 pigment) of particle size 50–60 μm at pigment concentration of 5 % exhibits a glow of 0.03 lux which increases to 0.04 lux as the concentration of the pigment is increased to 20 % in the print paste. Aqua blue pigment of particle size 10–15 μm (B1 pigment), the luminosity increases from 0.05 to 0.17 lux as the concentration of the pigment is increased from 5 to 20 % in the print paste. Similarly, for the pigment yellow-green (G1, G4 pigment) the lux readings change from (0.05 lux, 0.02 lux) at 5 % concentration to (0.15, 0.06 lux) respectively, at 20 % concentration. It was thus observed that as the concentration of pigment increases in the print, photoluminescence increases evidently irrespective of the size of the pigment. Similar studies have been reported in the literature by Zang et al. as well

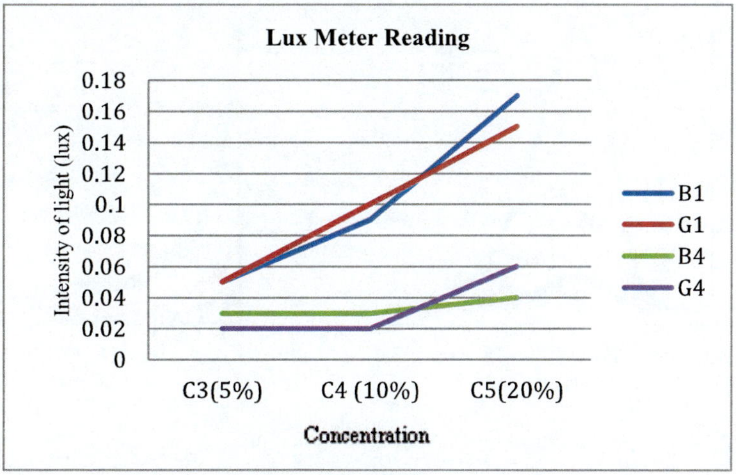

Fig. 2 Lux meter readings at a variable concentration

[11]. The lux reading of the photoluminescent printed samples were in the range of 0.02–0.17 lux, which is equivalent to starry night [2].

It was also observed that particle size of 10–15 μm for both aqua blue and yellow-green pigments (B1, G1) are exhibiting greater luminosity of (0.17, 0.15 lux) respectively in comparison to (B4, G4) of 50–60 μm pigment particle size exhibiting luminosity readings of (0.04 lux, 0.06 lux) respectively at 20 % concentration in the print paste. This may be attributed to the packing volume that plays an important role in the luminosity of the photoluminescent pigment printed swatches. It has been reported by Hom that lower particle size shall have the best packing volume and thus the best results [12]. Hom further elaborates that the bigger particles by their orientation scatter the emitted light in an undesired direction that reduces the emission intensity. Therefore, smaller particle size may perform better in luminous intensity. This has also been observed in the TRPL spectral scans hereafter.

Figure 3 illustrates the spectral scans for yellow-green pigment G4 of pigment particle size 50–60 μm at variable concentrations of C3 (5 %), C4 (10 %) and C5 (20 %) respectively as obtained from Time-Resolved Photoluminescence (TR-PL) spectroscopy. The figure clearly shows that the intensity of the luminosity of the pigment increases from 40 to 300au with an increase in concentration from C3 (5 %) to C5 (20 %) in the print paste. The decay profile of the photoluminescence shows an exponential curve with maximum intensity at the start of about 300 au at C5(20 %) concentration which gradually reduces to a level of 50 au which sustains for long-running parallel to the horizontal axis denoting time. The high initial brightness is owing to fluorescence in addition to phosphorescence at the beginning which attributes to the exponential rise at the start [5]. As the best performance was rated at 20 % concentration, the curtains for the elderly were printed at the same concentration in equal broad stripes with pigment B1, B4, G1 and G4 running lengthways.

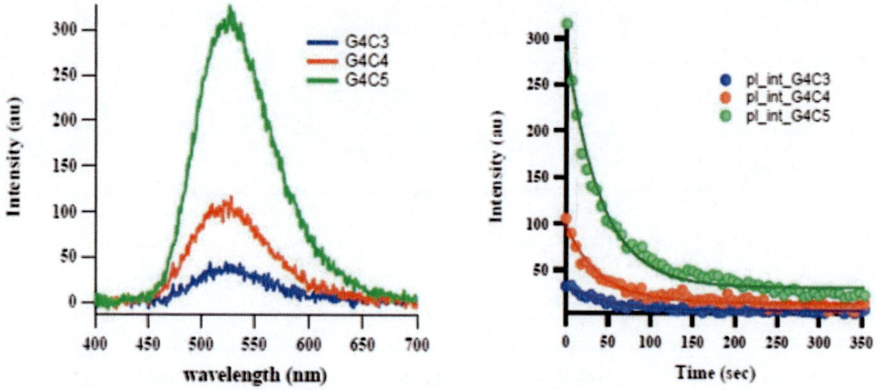

Fig. 3 Spectral scans of pigment G4 at variable concentration

4.2 *Effect of Different Types of Photoluminescent Pigment Printed Textiles on Luminescence*

In addition to cotton fabric, as mulberry silk fabrics and coarse silk fabrics like dupion and matka are also used in home furnishings, this section presents the effect of printing of PL pigment on coarse-textured silk fabrics. The pigment size and concentration were kept constant in all the fabric samples to observe the effect on the luminescence. It was observed that the maximum luminosity of the mulberry silk fabric was significantly lower in value (0.02 lux) in comparison with dupion (0.04 lux) and matka silk (0.04 lux) fabrics. To understand this phenomenon, SEM studies were conducted to study the surface morphology of the different fabrics. The SEM images in Fig. 4. demonstrate the surface morphology of cotton fabric before (control sample) and after printing. The samples were printed with blue pigment B4 of 50–60 μm at 20 % concentration. The deposition of the photoluminescent

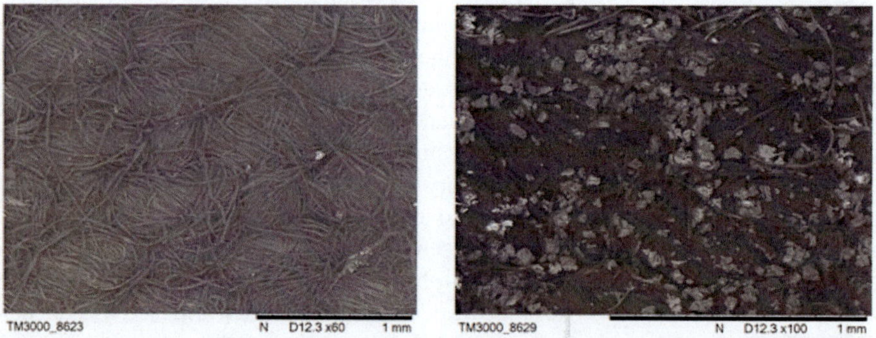

Fig. 4 Control Fabric Cotton (left) and fabric printed with Photoluminescent Pigment blue pigment B4 of 50–60 μm at 20 % (right)

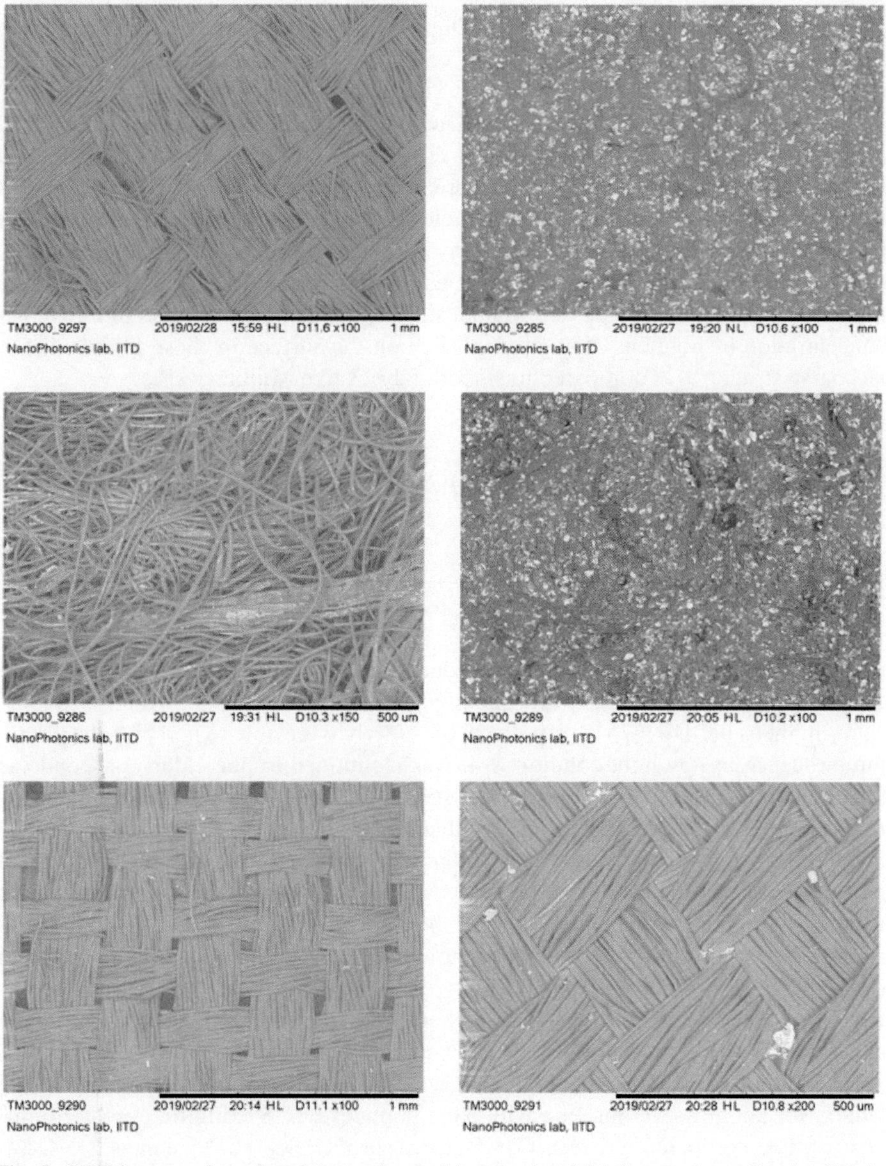

Fig. 5 SEM images of unprinted (control sample) and printed fabric samples. **a** Dupion Fabric, **b** Matka, **c** Mulberry

pigments as aggregates mostly at the intersection of warp and weft can be clearly seen on the cotton printed sample in Fig. 4. The overall distribution of pigment on cotton fabric is uniform.

Figure 5 presents the SEM images of Dupion silk (A), Matka silk (B) and Mulberry silk fabric which shows the deposition of the photoluminescent pigments on the surfaces of Dupion, Matka and Mulberry silk fabrics.

It was interesting to observe that the Dupion silk fabric (made from silk obtained from double cocoons, fabric with a coarse texture and rough feel) and Matka silk fabric (made from damaged cocoons, fabric with a coarse texture and rough feel) has multiple photoluminescent pigment particles deposited on the surfaces as compared to the Mulberry silk fabric (made from defect-free cocoons, with smooth fabric texture) with few photoluminescent pigment particles on the surface. Therefore, the higher luminosity of Dupion and Matlka silk fabrics can be attributed to the higher concentration of pigments that get deposited on the surface of these fabrics due to its coarse structure, as compared to smooth fabrics like Mulberry silk.

4.3 Visual Perception Study of Photoluminescent Pigments by the Elderly

The curtains were placed in the 10 houses of the elderly who volunteered for the visual perception study in their bedrooms. The interviews were taken the following morning and the user perception of the elderly and summarized. It was observed that even though the lux values of the printed swatches were below 1lux which is equal to starry night as per DIN67510 Standards, the elderly were able to perceive the photoluminescence in low light conditions. It was mentioned by the elderly respondents that the photoluminescence glow of the pigment is very soft, without glare and does not hurt the eye. The elderly also noted that in the night if their sleep was disturbed the nightlight may keep them awake. But now, with the photoluminescent aqua blue glow, the sleep of the elderly respondents was not disturbed even if the respondent woke up during nighttime. Irrespective of the level of disability and visual limitations of the elderly, all the four bands of photoluminescent pigments were visible, with (B1C5, G1C5) exhibiting better intensity of (0.15 lux, 0.17 lux) in comparison to B4C5 and G4C5. Table 2 gives a comprehensive analysis of pigment correlating the experimental readings with the visuals perception readings.

The photoluminescent pigment bands were visible distinctly as perceived by the elderly for about 2–3 h duration with Blue pigment B1C5 exhibiting greater luminosity when compared to green G1C5 at the particle size 10–15 μm. After 4–5 h duration, the elderly describe the photoluminescence on the curtain as a soft uniform glow. The individual bands seem to disappear. Out of 10 participants, 5 elderly homes experienced interference of corridor and light trespass.

Due to the interference of secondary lighting, the effect of the photoluminescent print was minimized. Two of the homes, the curtains were placed in the dark with no light trespass and these elderly users have given reviews to the scale of 4 out of 5. Edge definition and path-finding were rated best in these real-time dark spaces.

Table 2 Studies on user perception of photoluminescent printed textiles

S. No	Code	Concentration (20 %)	Lux	Au	User Perception	Broad Areas of Application
1	B1	C5	0.17	900	*****	Edge definition, nightlight, aesthetic features
2	B4	C5	0.04	210	*	Edge definition, aesthetic features
3	G1	C5	0.15	800	****	Edge definition, aesthetic features
4	G4	C5	0.06	300	**	Aesthetic features

*denotes least value ***** denotes highest value in the five point scale

An additional experiment was set up in the homes with light trespass. The window glass was covered with the black sheet of paper to eliminate the external light interfering with the photoluminescent print effects. The research findings show that the photoluminescent glow time was reduced to 2–3 h. This may be attributed to reduced excitation energy from daylight and also absorption of light by the black paper.

The average response of the elderly for the luminosity of the photoluminescent pigment printed textiles was rated in the range of 3–4 out of five. The pigment print B1C5 and G1C5 of 10–15 μm particle size were the best performers in the range of 4–5 stars. These pigments may find end use application in the area of path lighting and also emergency lighting. As the effect of light seems to start wearing off at 2–3 h, a short duration excitation light source may be programmed to rejuvenate the pigments to increase the performance of these photoluminescent pigments.

The photoluminescent pigment prints B4C5 and G4C5 with particle size 50–60 μm were rated in the range of 1–2 out of five for visible perception by the elderly. The usage of higher pigment size may not be effective for path lighting and navigation but may be used for aesthetics and personal spaces. The effect of the print is perceived best in a dark environment with no trespass of light from other sources. Hence it may be concluded that the prints have a potential to play a significant role in path lighting especially in a dark environment with a natural nightlight. Places with low levels of light trespass can also use the photoluminescent prints for edge definition say, for defining the edge of the carpet or furniture to prevent missteps and fall.

5 Design Development of Photoluminescence Pigment Printed Textiles

User preference study as an exploratory approach was conducted to assess the potential of photoluminescent pigment printed textiles as end products. Users were clearly able to navigate even at low pigment concentration in low light conditions. Figure 6 illustrates the various design possibilities with the print such as floral, geometric and

Fig. 6 Single screen prints under **a** photopic, **b** scotopic light

textural effects. The prints may be used as door handle covers for bathroom doors to help in navigation. Cushions, table runners, carpet and area rugs, edge of bedding, curtains, etc., maybe printed with photoluminescent pigments to facilitate navigation and path lighting in home spaces.

Figure 7 illustrates how a room may be designed using photoluminescent prints on products used on an everyday basis by the elderly to facilitate navigation. The flexibility of textiles may provide elderly freedom of placement and a higher degree of personalization for engineered design.

Fig. 7 Bedroom set with window treatment, carpet, pillows and coverlet **a** photopic, **b** scotopic light

6 Conclusion

It may be concluded that in photoluminescent pigments aqua blue and yellow-green pigment size, concentration and surface texture play a significant role, in the intensity of luminescence exhibited by the photoluminescent pigment printed textiles. Pigments aqua blue and yellow-green at the particle size of 10–15 μm were visually perceived best by the elderly at the photoluminescent pigment concentration of 20 % of print paste for an average duration of 4.5 h. The average lux reading observed was about 0.16 lux on cotton fabrics which is equal to starry night as per DIN67510 Standards. Dupion and Matka silk printed fabrics exhibited higher luminosity than mulberry silk printed fabrics. This may be attributed to the coarse texture of these fabrics which supported in the deposition of more PL pigments during printing than the smooth fabric surface of mulberry silk fabric as per the SEM studies.

Though the lux value of the PL pigment printed fabrics was recorded below 1 lux, the luminosity of the PL printed samples is visually perceptible by all the users. The elderly were able to perceive the photoluminescence of the photoluminescent pigments aqua blue and yellow-green pigment in low nightlight conditions. It has been perceived by the elderly that the photoluminescence exhibits a soft glow with no glare. The luminosity does not cause disturbances in the sleep patterns of the elderly.

Studies on real nighttime environment showed that a diverse range of products in home fashion may be designed using photoluminescent pigment printed textiles.

Therefore photoluminescent pigments in print in the concentration of 20 % and particle size of 10–15 μm may find applications in the design of interiors and soft furnishings at home for the elderly. Photoluminescent printed fabrics may be used as nightlight using for the elderly for safety and preventive measure. It may additionally be used for edge definition and path lighting. The prints may be used as nightlight, door handle covers for bathroom doors, refrigerators, etc., to help in navigation at night. Cushions, table runners, carpet and area rugs, edge of bedding, curtains, etc., maybe printed with photoluminescent pigments to facilitate navigation and path lighting in home spaces. Thus aqua glow from these photoluminescent printed textiles may play a key role in the design of products for the home to optimize cognitive performance especially to facilitate easy and safe navigation in a dark environment for the elderly.

Acknowledgements I would like to acknowledge the support of Dr. G. Vijaya Prakash, Nanophotonics Lab, Department of Physics, Indian Institute of Technology-Delhi, Hauz Khas, New Delhi, 110016, India for providing lab & support for Time-Resolved Photoluminescence Spectral Studies.

References

1. Philippe E, Gravisse JFVS (1997) United States Patent [19]. US; 19, pp 1–4
2. Gulrajani BML, Agarwal A, Bajaj A, Gupta A, Lohia C, Goel D (2008) Self- illuminated safety jackets. Fiber to fashion e-magazine [Internet]. https://www.fibre2fashion.com/industry-article/12/1187/self-illuminatedsafety-%0Djackets2.asp. Accessed 31 Jan 2016
3. Diana Friedrich lrsee (DE) (2013) United States, Pub. No: US 2010/0041620 A1
4. Daniel E, Zampa C, Alex A, Szopinski DP (1995) United States Patent no 5,415,911, pp 0–4
5. Yan Y, Zhu Y, Guo X, Ge M (2014) The effects of inorganic pigments on the luminescent properties of colored luminous fiber. Text Res J 84(8):785–792
6. Sharma R, Bairagi N (2019) Photoluminescent printed fabrics: an innovative solution to natural nightlight. In: Majumdar A, Gupta D, Gupta S (eds) Functional textiles and clothing. Springer Singapore, Singapore, pp 107–117
7. Vandewalle G, Maquet P, Dijk DJ (2009) Light as a modulator of cognitive brain function. Trends Cogn Sci 13(10):429–438
8. Cajochen C, Frey S, Anders D, Späti J, Bues M, Pross A et al (2011) Evening exposure to a light-emitting diodes (LED)-backlit computer screen affects circadian physiology and cognitive performance. J Appl Physiol 11;110:1432–1438
9. Australian Government Department of Health (2017) What causes falls? [Internet]. health direct. 2015. https://www.healthdirect.gov.au/what-causes-falls. Accessed 24 Aug 2017
10. Fuller GF, House W, Clinic M (2000) Falls in the elderly. Am Fam Physician 61(7):2159–2168. https://www.aafp.org/afp/2000/0401/p2159.html. Accessed 24 Aug 2017
11. Zhang X, Zhang X, Zhang J, Wang M, Zhao H, Lu S et al (2007) Size manipulated photoluminescence and phosphorescence in CaTiO 3:Pr3+ nanoparticles. J Phys Chem C 111(49):18044–18048
12. Hom NL (2010) Preparation and Properties of Long Persistent Sr 4 Al 14 O 25 Phosphors activated by rare earth metal ions. PhD thesis. Department of Energy and Materials Science Graduate School of Science and Engineering, Saga University, 145 pp.

Statistical Optimization of Ammonium Sulfamate and Urea-Based Fire Protective Finishing of Jute Fabric

Ashis Kumar Samanta, Reetuparna Bhattacharyay (Roy), Arindam Bagchi, and Ranjana Chowdhuri

1 Introduction

Some fundamental and preliminary studies on the thermal behavior [1–3] of jute and fire retardant finishing [4–7] of jute have been reported earlier [1–7]. Since long, Cellulosic fabrics are being made fire-resistant by treatment with different known non-durable and durable fire-retardant formulations as mentioned in literature and a comprehensive review by Horrocks and also by others where categorically they have mentioned different Fire Retardant Compounds and their efficacy, etc., in their research literature [8–11], including many reports on fire protective finishing of cotton, but the similar study on jute is very much limited and scanty [4–7]. Some preliminary studies on temporary fire retardancy of jute have been also reported earlier by Banerjee et al. [4] using borax-boric acid and di-ammonium phosphate. In a review described by Samanta [5] for different approaches of chemical finishing of jute, the use of potassium-sodium-tartrate (Rochelle Salt) as a fire retardant agent for jute fabric is mentioned.

Only a few scanty reports [4–7] are available in the literature on the application of fire retardant finish on jute, which warrants for an integrated study on this aspect with particular reference to jute that too with eco-friendly chemicals in today's context. Moreover, the recipe and conditions for the chemical processing and

A. K. Samanta (✉) · R. Bhattacharyay (Roy) · A. Bagchi
Department of Jute and Fibre Technology, Institute of Jute Technology, University of Calcutta, Kolkata 700 019, West Bengal, India
e-mail: ijtaksamanta@hotmail.com

R. Chowdhuri
Chemical Engineering, Department, Jadavpur University, Raja Subodh Mallik Road, Kolkata 700 032, West Bengal, India

A. Majumdar et al. (eds.), *Functional Textiles and Clothing 2020*,
https://doi.org/10.1007/978-981-15-9376-5_9

finishing of cotton cellulose are not exactly applicable for similar chemical modification of jute products. Hence unlike cotton, any chemical processing of jute has to be optimized/standardized separately.

Since the last decade, extensive research has been going on by using the new eco-friendly chemical formulation for improving fire retardant (FR) property on cellulosic and even on synthetics. However, the application of halogen-based fire-retardant is being greatly limited due to the possible production of dioxin and halogen gas emission with a large amount of smoke generation and corrosive gases evolution during the combustion of the said halogen-based fire-retardant materials. Most of the commercially available phosphorous-based fire retardant compounds are also susceptible to toxicity problem due to formaldehyde release problem from THPC, for example, Tetra-kis hydroxyl-methyl phosphonium chloride-based fire retardant formulations. Some fire retardant agents are rejected for not being wash fast or produce colors or for cost aspects.

Today, the application of fire protective jute textiles has widened in different fields such as cover fabrics for the temporary structure, brattice cloth in mines and also other decorative and furnishing applications such as car seat cover, curtains and furnishing fabrics used in public places such as school, colleges, hospital and theatre/cinema and conference halls besides as floor coverings, floor mats and carpets, etc. Different fire retardant chemicals act in a different way by manipulating pyrolysis, prevention of flammable gases, less or more char formation, controlling combustion, protecting from heat/oxygen and fuel, etc. Recently, formaldehyde-free fire-retardant finishing has been encouraged for environmental concern [12]. Synthesis of nano-ZnO and use of nano ZnO powder for fire protective finish of jute fabrics by applying nano-ZnO Powder admixed with poly-hydroxy amino siliconate binder have been studied recently [13, 14]. However, to avoid toxic fire-retardant chemicals, extensive research has been going on now to develop newer eco-friendly fire-retardant formulations, which are to be semi-durable or durable as per end uses and must be cost-effective also.

Sulfur-based fire retardants such as sulfamic acid and sulfamate compounds were earlier used in long past for developing the fire retardancy of cotton, paper and wood, but somehow these were dropped for lesser durability and perhaps for some yellowing problem, which, however, can be improved now with latest advanced techniques and knowledge on the same. Sulfamates have been used for the synthesis of many antibiotics, anti-cancer drugs, antiseptic drugs and weight loss drugs and other medical uses, besides its possible newer revived application in fire-retardant finishing of textiles. Ammonium sulfamate decomposes at elevated temperatures and releases some non-combustible gases, such as NH_3 and H_2O, etc., which reduces fire, besides chemical reactions between jute-cellulose and ammonium sulfamate at appropriate reaction-conditions. Some newer reports [15–21] on the use of ammonium sulfamate (AS) combined with other agents have been cited to improve fire retardancy for PA6 (Nylon 6), polypropylene and cotton, etc., which encourages the present authors to work on it for developing low cost and eco-friendly fire retardant recipe for jute fabrics to achieve at least LOI-value of 32–35 or above.

Considering all these as background information, in the present study, an attempt is made, to find newer eco-friendly fire-retardant formulations for jute fabrics using low cost indigenous common chemicals such as ammonium sulfamate (AS)/sulfamic acid (SA) along with urea to develop a fire protective jute fabric for its uses as cover fabric for a temporary structure, brattice cloth in mines and few furnishing and decorative purposes. The fire-retardant performance of each formulation is investigated for different concentration and combination in terms of LOI value, fire-spread time, afterglow time, char length and changes/loss in fabric tenacity, bending length in the said preliminary study. Thereafter, based on results obtained in the said preliminary study, a specific statistical optimization is carried out by using the User-Defined Quadratic Model (UDQM) technique to predict the optimum results for the maximum level of LOI, the minimum level of char length and loss in fabric tenacity as three important resultant variables, against two major independent input variables, that is the concentration of ammonium sulfamate (AS) and concentration of urea for obtaining suitable response surface equations. For analysis of reaction mechanisms with functional groups content/pattern, the study of FTIR spectra is also analyzed and reported.

2 Materials and Methods

2.1 Fabrics and Chemicals

The fabric used in this work was –6 % H_2O_2 (50 %) bleached fine hessian quality plain weave jute fabric with 60 ± 2 ends/dm, 60 ± 2 picks/dm, having fabric areal density of 220 ± 5 g/m^2. Warp count was 280 ± 5 tex (average 8 lbs/spyndle jute count) and weft count was 350 ± 5 tex (average 10 lbs/spyndle jute count). The said jute fabric had 0.80 ± 0.02 mm thickness and the initial whiteness index was 62 ± 2 (Hunter lab scale).

Laboratory grade ammonium sulfamate and sulfamic acid (From Loba Chemicals, India), Laboratory grade urea (From Loba Chemicals, India), arkofix NEC Plus liquid (an acrylic poly-carboxylate reactive binder, from erstwhile Clariant India Pvt. Ltd), ceralube HD liquid [a polyolefin dispersion as softener from erstwhile Clariant India Pvt. Ltd), Laboratory grade magnesium chloride as catalyst (From Loba Chemicals, India), saraflam-CWF, a commercially available THPC (Tetra-kis-Hydroxy-methyl Phosphonium Chloride)-based fire retardant agent obtained from Sarex Chemicals, Mumbai, India were used in this work.

2.2 *Fire Retardant Finishing Treatment*

Bleached control jute fabrics were padded (100 % weight pick-up) with the following fire retardant formulations by padding (2-dip and 2 nip) followed by drying at 100 °C for 10 min and curing at 130 °C for 3–4 min.

Sl. no	Type of formulation used in varying concentration	Formulation code used
1	Urea + Ammonium Sulfamate + Magnesium Chloride (as catalyst)	J-U-AS
2	Urea + Sulfamic Acid + Magnesium Chloride (as catalyst)	J-U-SA
3	Saraflam-CWF (Commercial THPC-based Fire-retardant agent) + Magnesium Chloride (as catalyst)	J-S-CWF
4	Urea + Ammonium Sulfamate + Arkofix NEC Plus binder + Ceralube HD softener + Magnesium Chloride (as catalyst)	J-U-AS-A-C

Above said combination of fire retardant finishing treatments of jute fabric with the above formulations 1–4 were applied on bleached jute fabric in varying concentrations for obtaining an optimized formulation based on results of fire protective performance properties and fabric tenacity tested before and after treating the bleached jute fabric separately and also in combination as a mixture with varying concentration levels of the selective chemicals at the same and comparable conditions by above said pad-dry-cure technique, with or without the use of pre-fixed dosages of arkofix-NEC-an acrylic poly-carboxylate binder and ceralube-HD as a polyolefin-based softener emulsion, always with very low dosages of an additional acidic catalyst (magnesium chloride) for promoting the reaction of AS with jute-celluloses to improve its wash fastness.

2.3 *Conditioning for Physical and Fire Retardant Performance Tests*

All the selected untreated and treated jute fabric samples were conditioned by exposure to an atmosphere of 65 ± 2 relative humidity for at least 24 h [22] before carrying out all tests as mentioned below.

2.4 *Test of Tensile Properties*

Warp-way breaking tenacity (cN/tex) of jute fabrics was determined by the raveled strip method as per IS-1969–1968 [22] using an Instron (Model-1445) CRT-Universal tensile strength tester. Traverse speed was set at 100 mm/min using a pretension of 0.5 N. The gauge length of the jute fabric samples was 50 mm with the cut sample dimension of 100 mm length × 20 mm width after raveled from two sides. Test results reported was an average of 10 results for each case.

Fabric tenacity was determined by the following calculations:

$$\frac{Breaking\ Load(cN)}{Fabric\ wt\ \left(gm/mt^2\right) \times Fabric\ Width\ (mm)} \quad (1)$$

The retention % of the tensile strength or tenacity for selected treated and untreated jute fabric specimen tested was calculated using the following calculations:

$$\frac{Treated\ fabric\ Tenacity}{Untreated\ fabric\ Tenacity} \times 100 \quad (2)$$

2.5 Evaluation of Bending Length

The bending length of selective jute-based fabric samples was measured following IS-6490–1971 method [22] with the standard specimen size of 200 mm × 25 mm. Higher is the bending length, stiffer is the fabric and vice -versa. The test results was obtained by an average of using five tests for each sample.

2.6 Determination of Limiting (Critical) Oxygen Index (LOI) Values

Limiting (Critical) Oxygen Index (LOI) value indicates the relative measure of fire retardancy performance. A textile fabric/material is said to be fire retardant, if LOI value of the said textile material is above 27. LOI value is determined as the minimum value of volume concentration of O_2 (in a mixture of O_2 and N_2), to sustain the fire for combustion of the textile/similar related material and below which level of volume concentration of oxygen, the fire will extinguish. Critical/limiting oxygen index (LOI) values of untreated and treated jute fabric samples were determined following ASTM-D-2863–77 method [23] as per the calculation given below:

$$\text{Critical Oxygen Index (n)} = \frac{\text{Volume concentration of } O_2 \text{only}}{\text{Volume concentration of } (N_2 + O_2)} \times 100 \quad (3)$$

2.7 Flammability Test

The jute fabric samples cut in size of 2.75 inch wide × 10.0 inch long, were mounted in a suitable clamp in a standard vertical flammability tester made by MAG-Solvic –Coimbatore, India. The bottom edge of the fabric was then subjected to burning using a standard flame for exactly 10 s time. After removal of the flame, the flame spread time or flame extinguishing time, char length and afterglow time were noted using a stopwatch and scale. This test was done as per ASTM D6413/D6413M-13b standard test method [24].

2.8 FTIR Analysis

Selected samples of both untreated and treated jute fabrics were repeatedly washed in water followed by washing by ethyl alcohol and acetone before taken for analysis for FTIR spectra as per usual KBr pallete method using Perkin Elmer make SPECTRUM-II model of FTIR.

2.9 Method of Optimization of Fire-Retardant Formulation Using Response Surface Methodology

Response surface methodology (RSM) is a useful statistical tool for understanding and optimizing interactive effects amongst different process variables for optimizing the process conditions. Response surface methodology (RSM) thus determines interactive effects between selective experimental input variables for obtaining selective output response/resultant variables by using suitable statistical design of experiment model such as UDQM (user-defined quadratic model). This UDQM model [25] was used earlier for optimization of fire retardant formulations by application of nano-ZnO powder along with poly-hydroxy amino siliconate binder on jute. Based on experimental results of the preliminary study, the average dosages of fire-retardant chemicals (ammonium sulfamate and urea) and numbers of test run for UDQM model of experimental design set up under fixed conditions of treatment are determined for optimizing these two most important input process variables (concentration of ammonium sulfamate and concentration of urea) for maximizing limiting oxygen Index (LOI) value, minimizing char length and loss of fabric tenacity for corresponding fire retardant treatment of jute fabric, for optimum results.

As per UDQM model of statistical experimental design, total numbers of test run required (15) for optimizing 2 input variables, that is, x_i (concentration of urea, i.e., x_1 or A and concentration of ammonium sulfamate, i.e., x_2 or B) at 5 levels (−2, −1, 0, + 1, + 2) to determine three resultant response variables (Yi), that is average LOI value (Y1 or R1), char length in mm (Y2 or R2) and loss of fabric tenacity %

(Y3 or R3), which can be represented by the following quadratic response surface equation, where, this RSM equation can predict each of the respective resultant response variables (Yi or Ri) against specific input variables (x_i):

$$Yi\ or\ Ri = b_o + \sum_{i=1}^{i=k} b_i x_i + \sum_{i=1}^{i=k} b_{ii} x_i^2 + \sum_{i=1}^{i=k} b_{ij} x_i x_j \quad (4)$$

where b_o, b_i, b_{ii} and b_{ij} are corresponding regression coefficients, to be obtained from the analysis of variations data.

3 Results and Discussion

3.1 *Preliminary Study on Fire Retardant Performance of Bleached Jute Fabric Treated with Urea and Ammonium Sulfamate/Sulfamic Acid Individually and in Suitable Combinations*

The main aim of this present work is to find a low-cost indigenous chemical formulation as optimized.

Recipe for fire-retardant finish of jute fabric using common eco-friendly fire retardant chemicals such as ammonium sulfamate (AS) and urea. Based on earlier reports of fire retardant finishing treatment of nylon-polyamide and cotton-cellulose using sulfamate compounds [15–21], it is thought that sulfamate-based compounds may also be equally effective for fire retardant treatment for ligno-cellulosic materials such as jute. So, sulfamic acid or ammonium sulfamate and urea have been applied on jute (a ligno-cellulosic fibre having 58–64 % alpha–cellulose content) individually and in their varying combinations to achieve an acceptable level of fire retardant property on jute fabrics. Hence, in a preliminary study, bleached jute fabric was separately treated with 5–20 % sulfamic acid, 5–20 % ammonium sulfamate and 5–20 % urea individually and their different combinations to obtain a certain level of fire retardant property with minimum loss of fabric tenacity and a good and acceptable level of fire retardant property. The relevant results are shown in Tables 1 and 2.

Relevant results in Table 1 indicate that said fire retardant agent (sulfamic acid/ammonium sulfamate and urea) individually cannot achieve the required fire retardant performance of jute fabric and suffers considerable loss of fabric tensile strength (18–40 % loss in fabric tenacity for 5–20 % sulfamic acid treatment and 16–30 % loss in fabric tenacity for 5–20 % ammonium sulfamate treatment). The loss of tensile strength is presumed to be due to acidic degradation of jute fabric (in situ acidic environment generated from either sulfamic acid or from ammonium sulfamate and also additional acidic nature of magnesium chloride catalyst used).

Table 1 Fire retardant performance of bleached jute fabric treated with Ammonium Sulfamate or Sulfamic Acid and Urea Individually

Treatments*	Physical properties				Vertical flammability test			LOI %
	% Loss in tenacity	Elongation (%)	Bending Length (cm)	Whiteness index (Hunter)	Fire spread time (s)	After glow time (sec)	Char length (cm)	
Control bleached jute	–	3.4	4. 0	62 ± 2.0	50(BEL*)	55	–	20
Sulfamic Acid 5 %	18 ± 3.0	3.5	4.5	59 ± 2.2	32 SE *	40	8.5	21
Sulfamic Acid 10 %	23 ± 3.0	3.7	4.5	58 ± 2.1	30 SE	33	7.0	25
Sulfamic Acid 15 %	30 ± 4.0	3.9	4.6	57 ± 2.4	23 SE	28	6.0	27
Sulfamic Acid 20 %	38 ± 3.5	4.1	4.2	58 ± 2.2	21 SE	25	5.0	30
Ammonium Sulfamate 5 %	17 ± 2.5	3.3	4.5	59 ± 2.6	28 SE	38	6.0	23
Ammonium Sulfamate 10 %	20 ± 2.8	3.4	4.3	57 ± 3.0	22 SE	30	5.0	28
Ammonium Sulfamate 15 %	24 ± 3.0	3.5	4.2	56 ± 3.1	15 SE	24	4.5	30
Ammonium Sulfamate 20 %	28 ± 3.0	4.2	4.6	55 ± 3.2	16 SE	20	4.0	32
Urea 5 %	05 ± 2.0	3.5	4.3	54 ± 3.2	55 (BEL)	–	BEL	20
Urea 10 %	5.5 ± 2.7	3.6	4.6	53 ± 3.1	52 (BEL)	–	BEL	21
Urea 12.5 %	08 ± 3.0	4.0	4.7	52 ± 3.0	54 (BEL)	–	BEL	22
Urea 15 %	09 ± 2.0	4.0	4.9	50 ± 3.0	56 (BEL)	–	BEL	23
Urea 20 %	11 ± 2.0	4.1	4.8	50 ± 3.4	60 (BEL)	–	BEL	22

*All treatments are done with 1/5th of $MgCl_2$ catalyst; BEL: burned entire length; S.E.: self-extinguishing

LOI value for application of 15 % to 20 % sulfamic acid only on jute is around 27–30, char length is 5–6 cm and there is 31–40 % loss in fabric tenacity. While, 15–20 % ammonium sulfamate treated jute fabric shows LOI value as 30–32 with 4–4.5 cm char length having 25–30 % loss in fabric tenacity. The higher loss in fabric tenacity for urea treatment may be viewed as an effect of acidic degradation by $MgCl_2$ catalyst used, not for urea (rather urea reduces some acidity in the treatment bath and hence the loss of fabric tenacity is much less). Thus, the use of more than 15 % sulfamic acid causes more than 30 % loss in fabric tenacity and the use of more than 15 % ammonium sulfamate causes more than 25 % loss in fabric tenacity. Irrespective of % application, only urea treatment does not confer any noticeable fire retardant property on jute fabric, though the loss of fabric tenacity is much less, which is obvious as urea as a nitrogenous agent acts as a reaction promoter for fire retardant chemicals such as phosphates and sulfamates for promoting the reaction of phosphates/sulfamates with cellulose, where urea individually is not acting as a fire-retarding agent at all. So, in the next set of preliminary experiment, a suitable binary mixture of ammonium sulfamate and urea or sulfamic acid and urea as a selective mixture in varying proportions were applied on bleached jute fabrics to achieve an acceptable level of fire-retardant action on it.

Understanding the superiority of ammonium sulfamate and urea combination for fire retardancy performance over sulfamic acid and urea admixture, ammonium sulfamate and urea combination is considered to be preferable for use as the main fire retardant agent for application on bleached jute fabric in presence of low dosage of $MgCl_2$ (1/5th of the main fire retardant chemical used) as the catalyst. It is also found that without $MgCl_2$, only sulfamic acid generated from ammonium sulfamate or during direct use of sulfamic acid, the reaction of sulfamate with cellulose was poor and hence fire retardant performance was lower, which may be due to less acidity and lesser acidic pH than required for such reaction to occur. So, the addition of $MgCl_2$ helped to promote the reaction between sulfamates and cellulose for blocking –CH_2OH group of cellulose at C6 position to reduce levoglucosan formation. The function of urea is considered to be manifold, for example it acts as a swelling agent for cellulose, act as positive supporting action to promote reaction potential between cellulose and phosphates or between cellulose and sulfamate s(as main fire retardant agent) by the contribution of lone pair of nitrogen of urea, and also provides some buffer action reducing acidity in the formulation, which reduces % loss in fabric tenacity.

Relevant results in Table 2 (for formulation 1 and 2 thus indicate that for formulation 1(a-g) with the application of 10–20 % ammonium sulfamate (AS) along with 10, 12.5 and 15 % urea in presence of 1/5th of $MgCl_2$ catalyst, LOI values are found to be nearly to 30–38 (while, LOI value for control bleached jute is 20 only) with 24–32 % loss in the tenacity of treated jute fabric. Results of this preliminary study finally show that the use of 12.5 % urea and 15 % ammonium sulphamate (AS) combination (formulation 1e) gives more or less better balance of fire-retardant performance associated with some loss of tenacity of jute fabric having the highest LOI value of 38 and minimum char length of 2 cm. Increase in urea from 12.5 to 15 % with 15 % AS do not further improve LOI, rather reduces it to 35 from 38, as obtained in formulation 1f, vide Table 2.

It may also be noted that treatment with 10 % urea and 15 % ammonium sulfamate (AS) combination (formulation 1b) in presence of said $MgCl_2$ catalyst shows almost nearer or at per results with formulation 1e, showing a bit lower LOI 34 and char length 3.5 cm associated with 26 % loss of fabric tenacity, whereas application of 10 % urea and 20 % AS combination (formulation 1d) treatment causes 27 % loss of tenacity of jute fabric with no further significant improvement in LOI value (34–34.5). With an increase in ammonium sulfamate (10–20 %), the increase in LOI value up to a certain level (30 to 38) may be explained by increased blocking of C_6 position of jute cellulose by sulfamates preventing the formation of levo-glucosan (the flammable gas produced for propagating fire). For approx. 10–15 % add on of AS, treated jute fabric may become somewhat stiffer and hence, checking of bending length, as an indicator of fabric stiffness (indirect indicator of fabric hand) is justified. After such fire retardant treatment, there is some increase in bending length values from 4.2 to 4.8 or so, Table 2, showing a very little increase in stiffness after such treatment, not spoiling the hand value of fabric.

As said earlier, for comparison purpose, *Saraflam-CWF (THPC-based),* a commercially available fire retardant agent was applied on the same bleached jute fabric under comparable conditions of treatment and corresponding results are also shown in Table 2. Comparison of relevant results of Formulations 1(a to g) with Formulation 3 (a to d) indicate that the fire retardant performances for treatment with 15 % ammonium sulfamate and 12.5 % urea with 3 % $MgCl_2$ are somewhat appear to be better in terms of LOI value (38 for 15 %AS + 12.5 % urea treatment) than LOI value obtained by the equal dosage of treatment with 15 % saraflam-CWF (formulation 3b showing LOI as 33) with a somewhat higher loss of fabric tenacity and higher char length.

Before proceeding to the characterization and analysis of detailed fire-retardant performances for the said treatments and its optimization, it is better to discuss the possible reaction mechanisms to be confirmed by FTIR.

3.2 Reaction-Mechanisms

The cellulose of jute may react with ammonium sulfamate at elevated temperature possibly in the following manner as discussed under Reaction scheme 1, 2 and 3. Ammonium sulfamate at high temperature may react simply with the primary hydroxyl group at C_6 position of jute cellulose as shown under Reaction scheme 1. The chemical reactions as shown in Reaction 1–3 are considered to be possible, when urea and ammonium sulfamate are applied together on jute cellulose in presence of $MgCl_2$ as an additional acidic catalyst by pad dry-cure process at specific conditions, that is, by drying at 100 °C for 10 min and curing at 130 °C for 3–4 min.

Table 2 Fire retardant performance of jute fabric treated with different fire retardant formulations 1–4

Treatments*	Physical properties				Vertical flammability test			LOI %
	% Loss in tenacity	Elongation (%)	Bending length (cm)	Whiteness index (Hunter)	Fire spread time (s)	After glow time (sec)	Char length (cm)	
Control bleached jute	-	3.40	4. 0	62 ± 2.0	50(BEL*)	55	-	20
Formulation 1; Code J-UAS (Urea & Ammonium Sulfamate)								
Formulation 1a U 10 %; AS 10 %	24 ± 2.0	4.10	4.2	59 ± 2.5	22, SE	30	5.0	30.0
Formulation 1b U 10 %; AS 15 %	**26** ± 2.0	**4.41**	**4.3**	**58 ± 3**	**20, SE**	**20**	**2.5**	**34.0**
Formulation 1c U 10 %; AS 20 %	30 ± 2.0	4.63	4.6	57 ± 2.6	10, SE	18	3.0	33.0
Formulation 1d U 15 %; AS 10 %	27 ± 2.0	4.51	4.7	55 ± 3.2	12, SE	17	2.8	34.5
Formulation 1e U 12.5 %; AS 15 %	**28** ± 2.0	**4.62**	**4.8**	**56** ± 2.4	**10, SE**	**12**	**2.0**	**38.0**
Formulation 1f U 15 %; AS 15 %	27 ± 2.0	4.42	4.7	54 ± 2.4	12, SE	15	2.5	35.0
Formulation 1 g U 15 %; AS 20 %	32 ± 2.0	4.83	4.9	58 ± 3.0	7,SE	17	3.1	37.5
Formulation 2; Code J-USA(Urea & Sulfamic Acid)								
Formulation 2a U 10 %; SA 5 %	24 ± 2.0	4.60	4.1	60 ± 2.5	28,SE	30	4.9	25.0
Formulation 2b U 10 %; SA 10 %	**30** ± 2.0	4.71	4.2	61 ± 2.2	24, SE	26	4.5	27.5

(continued)

Table 2 (continued)

Treatments*	Physical properties				Vertical flammability test			LOI %
	% Loss in tenacity	Elongation (%)	Bending length (cm)	Whiteness index (Hunter)	Fire spread time (s)	After glow time (sec)	Char length (cm)	
Formulation 2c U 10 %; SA 15 %	34 ± 2.0	4.81	4.3	63 ± 2.6	19, SE	21	4.5	30.1
Formulation 2d U 15 %; SA 5 %	26 ± 2.0	4.40	4.2	59 ± 2.4	13,SE	19	4.0	28.5
Formulation 2e U 15 %; SA 10 %	31 ± 2.0	4.80	4.4	61 ± 2.8	17, SE	15	3.6	29.5
Formulation 2f U 15 %; SA 15 %	**40** ± 2.0	**4.97**	**4.6**	**62 ± 3.1**	**16, SE**	**17**	**4.0**	**33.0**
Formulation 3; Code J-S-CWF(commercial Fire retardant chemical SARAFLAM-CWF)								
Formulation 3a Saraflam-CWF 10 %	20 ± 2.0	3.85	4.2	58 ± 2.2	26, SE	18	5.5	30.0
Formulation 3b Saraflam-CWF 15 %	**26** ± 2.0	**4.26**	**4.5**	**57 ± 3.0**	**18, SE**	**16**	**4.0**	**33.0**
Formulation 3c Saraflam-CWF 20 %	30 ± 2.0	4.55	4.8	56 ± 2.5	12, SE	15	3.0	35.5
Formulation 3d Saraflam-CWF 30 %	**33** ± 2.0	**4.67**	**5.2**	**55 ± 4.0**	**10,SE**	**15**	**3.0**	**38.0**
Formulation 4; Code J-UAS-AC(Using Commercial acrylic-poly-carboxylate binder with polyolefin softener)								

(continued)

Table 2 (continued)

Treatments*	Physical properties				Vertical flammability test			LOI %
	% Loss in tenacity	Elongation (%)	Bending length (cm)	Whiteness index (Hunter)	Fire spread time (s)	After glow time (sec)	Char length (cm)	
Formulation 4: Urea 12.5 %; AS-15 %+ Arkofix NEC 50gpl + Ceralube HD 30gpl	30 ± 2.0	4.1	5.3	54 ± **4.0**	4, SE	12	1.5	38.0

*All treatments are done with 1/5th of $MgCl_2$ catalyst; BEL: Burned entire length; S.E.: Self-extinguishing

Table 3 LOI values of differently fire retardant treated jute fabric after **Soap washing done under ISO-II method

Formulation Code	Treatment	LOI-value before soap washing	LOI* –value after Soap Washing**		
			1st wash	3rd wash	5th wash
J-U-AS (Formulation 1e)	Urea—12.5 %, Ammonium Sulfamate – 15 % + Magnesium Chloride as catalyst (1/5th of AS)	38	35	33	31
J-U-SA (Formulation 2f)	Urea –15 %, Sulfamic Acid – 15 % + Magnesium Chloride as catalyst (1/5th of SA)	33	30	28	26
J-S-CWF (Formulation 3b)	Saraflam CWF 15 % + Magnesium Chloride as catalyst (1/5th of Saraflam-CFW)	33	31	30	29
J-S-CWF (Formulation 3d)	Saraflam CWF 30 % + Magnesium Chloride as catalyst (1/5th of Saraflam-CFW)	38	36	34	32
J-UAS-A-C (Formulation 4)	**Urea (12.5 %) and Ammonium Sulfamate (15 %), Arkofix NEC Plus (50gpl) and Ceralubu HD (30gpl) and Magnesium Chloride as catalyst (1/5th of AS)**	**38**	**37**	**36**	**36**

* LOI values shown in Table 3 are not absolute data, but are averages of three sets of data with variation Of ± 2.0 to ± 2.5, Having Average Values as Shown. ** Soap Washing Was Done as Per IS-II Method

Reaction Scheme-1

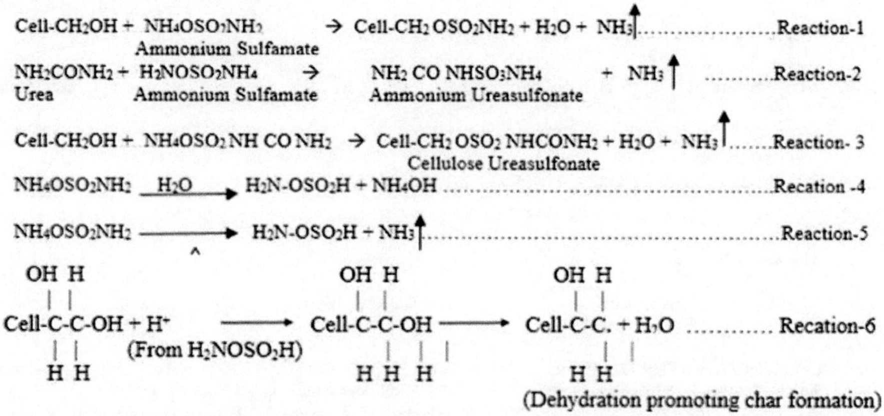

Reaction-3 in the above reaction—mechanism shows the blocking of C_6-CH_2OH groups by forming urea sulphonate complex. Moreover, a higher amount of AS produces a higher amount of sulphamic acid (SA) as per Reaction 4 H^+ ion (proton) produced from SA/AS (Reaction 5) ultimately promotes char formation of jute cellulose during heating/burning as shown in Reaction 5 and 6. Thus finally Reaction 3 and Reaction 6 improve fire retardant performance and increase the LOI value of the ammonium sulfamate-urea treated jute fabric samples.

Reaction scheme 2 shows that sulfamic acid may be hydrolyzed in the presence of water as shown in Reaction 7 and the hydrogen cation (H^+ ion) liberated from sulfamic acid then may follow the Reaction 6 in Reaction scheme 1 as above for promoting char formation in jute celluloses and hemicelluloses when heated and burns and became fire retardant to a Specific LOI level, as shown in Table 2.

Jute cellulose may react similarly with sulfamic acid (H_3NSO_3) at much higher and elevated temperature (>200 °C) as shown in Reaction-8 as per Reaction scheme 2, if the heating is done above the decomposition temperature of SA, that is above 205 °C, as 205 °C is the decomposition temperature of SA). But, in this case, as jute fabrics are heated at 130–140 °C that too for only 3–5 min, the Reaction 8 cannot occur. Similarly, the Reactions 9 and 10 (under Reaction scheme 3) also cannot occur, as the temperature is not above 205 °C, in any case. But when Urea and Ammonium Sulfamate combination is used Reactions 1–3 and 6 only may occur if AS + Urea treated jute fabric samples is heated during curing at a temperature of 130 °C for 3–4 min. Thus, the supremacy of ammonium sulfamate and urea combination is established here and understood well from the analysis of these reaction schemes.

Reaction Scheme-2

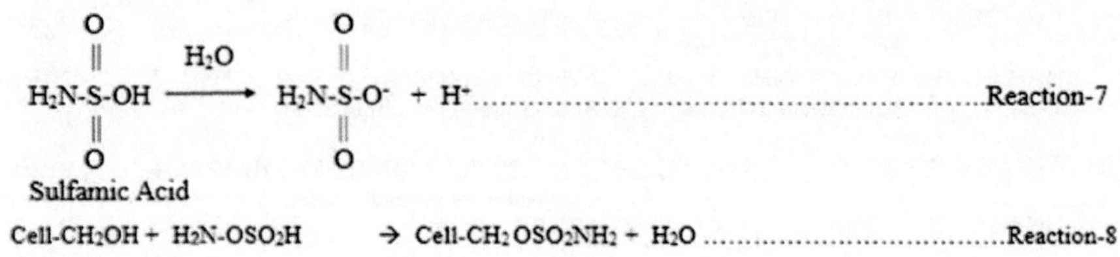

Reaction Scheme-3

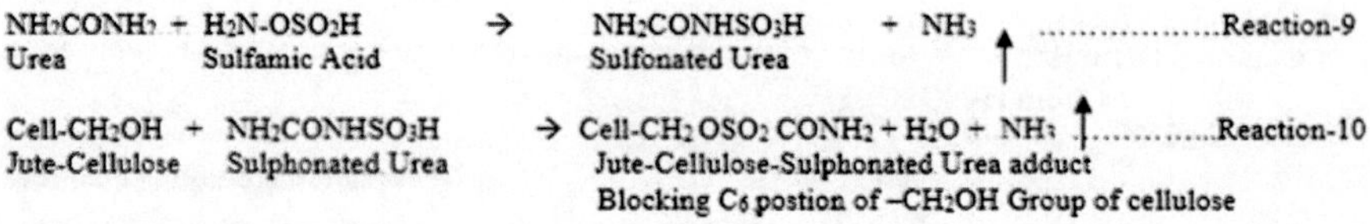

Thus it is evident from this preliminary experimental data observed that the fire retardant performance of 12.5 % urea and 15 % ammonium sulfamate (AS) treated jute fabric is found to be superior even as compared to the same or higher dosages commercially available/non-ecofriendly fire-retardant formulations (15 % saraflam-CWF) and is also better from similar dosages of treatment with urea and sulfamic acid treatment. However, to establish 15 % ammonium sulfamate and 12.5 % urea treatment with 3 % $MgCl_2$ as the catalyst to be useful fire-retardant formulation for jute, it is essential to analyze the wash stability of treated jute fabric samples and to understand changes in functional groups pattern by FTIR analysis before and after different fire retardant treatments before optimizing the said fire-retardant formulation using suitable statistical design of the experiment.

3.3 Test of Wash Stability of Differently Fire Retardant Treated Jute Fabrics

After normal washing of the said fire retardant treated jute fabrics with 5gpl nonionic detergent solution for 30 min at 30 °C, that is, at room temperature as per IS-II method of wash, the fire retardancy performance in terms of LOI values is found to be decreased to some extent after 1–5 cycles of washing and these results are shown in Table 3. It is observed that formulation 1e (J-U-AS – i.e., 15 % AS + 12.5 % urea with 3 % $MgCl_2$ catalyst) the reduction in LOI value per cycle is almost two to three unit per wash cycle, while for formulation 2f (J-U-SA) the reduction rate of LOI values are almost same, but after five wash cycles formulation 1e (J-U-AS) shows higher LOI as 31 while formulation 2f shows LOI as 26 after five wash cycles. Considering the loss of LOI values in washing up to five cycles, post-treatment with 50 gpl arkofix–NEC as acrylic–poly-carboxylate binder and 30gpl ceralube–HD (a polyolefin-based softener) after treatment of jute fabric with 15 % AS and 12.5 % urea + 3 % $MgCl_2$ (as a special wash fast recipe, i.e., formulation

4, as shown in Table 2) and this formulation 4 is when modified as formulation 1e with the admixture of 50 gpl arkofix NEC Plus binder with 30 gpl ceralube–HD softener show much higher wash stability of the formulation up to five wash cycles. Additional cross-linked acrylic binder film formed after drying and curing embedding said softener compound, fire retardant performances do not impair as these acrylic binders and polyethylene-based softener are not acting as a single compound after curing, but are forming a crosslinked acrylic binder film embedding the softener on the jute-cellulose-sulfamate adduct substrate after the said fire-retardant treatment. The reduction in LOI value per cycle is a maximum 1 unit per wash cycle up to third wash cycle and there is no further loss of LOI value after third to fifth wash cycles, in this case, showing good wash stability. formulation 1e and formulation 3d almost show equal fire retardant performance and wash stability, but formulation 3d need 30 % saraflam–CWF treatment, which is costlier and has an objectionable formaldehyde release problem, as it is a THPC-based fire retardant compound.

The higher loss of LOI value in all the cases (except formulation 4) is due to the removal of partial unfixed fire retardant chemicals applied and also for hydrolysis of some already fixed fire retardant chemical formulations, however, which is not so rapid after third to fifth wash because such unfixed chemicals are already washed out in first and second washes. However, the changes in LOI value observed in third and fifth wash is probably due to the hydrolysis of the already fixed fire retardant chemicals, which is, however, nil for formulation 4 (marked bold in Table 3).

3.4 FTIR Spectroscopic Analysis of Untreated and Fire Retardant Treated Jute Substrate

FTIR spectra of (a) Bleached untreated Jute (Bleached Control Jute); (b) J-U-AS-formulation 1e; (c) J-U-SA formulation 2f; (d) J-Sara-CFW formulation 3d and (e) (d) J-U-AS-AC formulation 4, are shown in *Spectra* (a) to (e) in Fig. 1 from wave no. 4000–500 cm^{-1}. For all the five FTIR spectra there are some common characteristic features due to three major constituents of jute namely alpha-cellulose, hemi-cellulose and lignin [1–3] and their inter-unit linkages,. Such common characteristic peaks/features of FTIR spectra for each of the samples of untreated and treated jute samples are shown in Table 4.

J-U-AS-formulation 1e treated jute [Spectrum (b), Fig. 1], by and large, retain similar characteristics with their FTIR spectra as that obtained for the bleached control jute, thereby showing almost no changes in major functional group pattern by the said treatment, but there are some minor decrease/increase or weakening of existing peaks with or without some additional small peak often observed. In Spectrum (b), Fig. 1, for J-U-AS-formulation 1e, for treated jute, there is a minor increase in band intensity at 1030 cm^{-1} (due to $–OSO_3–$ stretching vibration) and a small peaks at 1109 cm^{-1} (indicating sulfamation in $-CH_2OH$ group in C_6 position of jute cellulose) by the said AS-based fire retardant treatment on jute fabric.

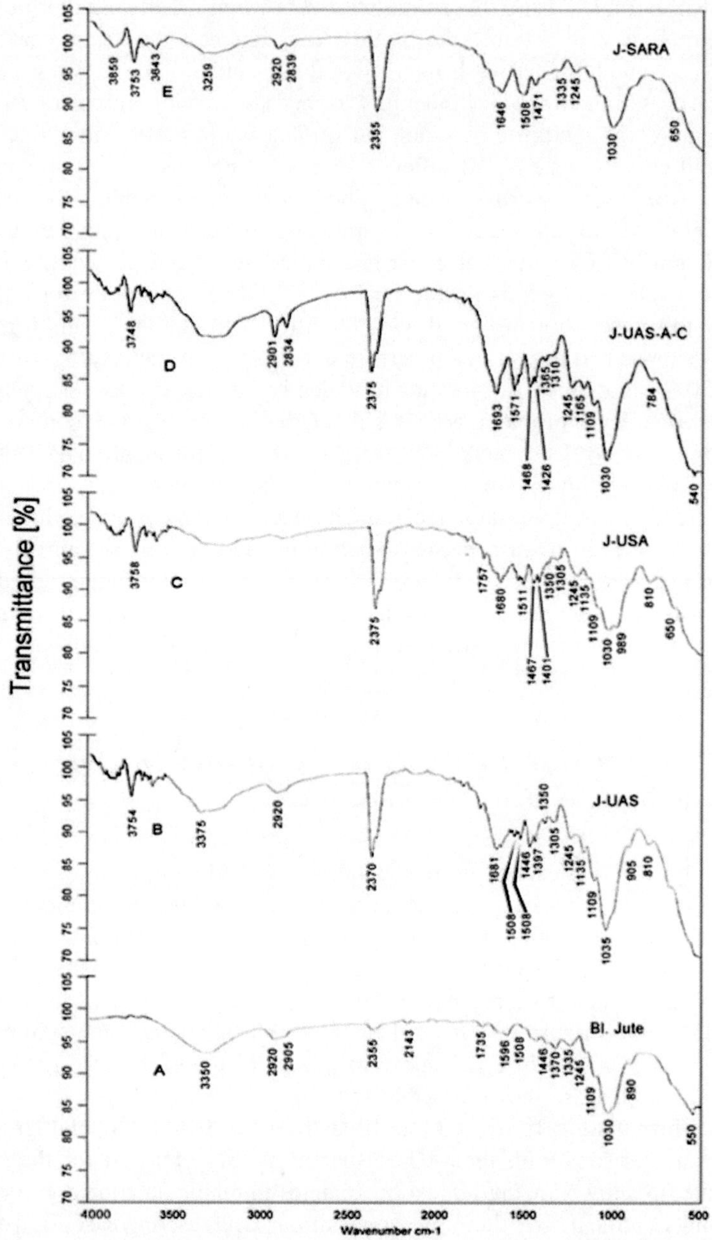

Fig. 1 FTIR Spectra of **a** Bleached Jute, **b** J-Urea + AS, **c** J-Urea + SA, **d** J-Urea + AS + Arcofix + Ceralube and **e** J-Sara-CWF treated Jute. Bl Jute- Bleached control Jute; J-U-AS-Formulation 1e; J-U-SA- Formulation 2f;; J-Sara-CWF-Formulation 3d; J-U-AS-A-C- Formulation 4

Table 4 Common FTIR peaks of untreated and fire retardant treated jute fabrics

Transmittance Peaks in FTIR Spectra (cm^{-1})	Significance of the Common FTIR peaks in different wave numbers
550–560	For –OH out of plane bending in celluloses of jute
880–890	For β-glycosidic linkage of cellulose chains in jute cellulose/hemicelluloses
1020–1030	Assigned to aromatic -C-H in plane deformation & -CH-OH deformation for primary alcohol in lignin
1240–1250	For –C–O–C and –C=O stretching indicating ether linkage as side substituent in xylan and also lignin aromatic -C=O stretching
1330–1340	For –OH in plane bending in C_6 –CH_2OH group in cellulose
1360–1370	For C-H deforming (symmetric), attributed to cellulose, hemicellulose and lignin
1440–145	For C-H deforming (asymmetric), attributed to lignin and other constituents of jute
1500–1516	For lignin aromatic ring vibration and stretching for jute lignin
1590–1610	For C=C stretching in lignin olefinic units present in jute
1720–1750	For C=O stretching of ester group in hemicelluloses of jute
2340–2350	For C–H stretching in polysaccharide chains of cellulose macromolecules
2900–2930	For C–H stretching vibrations of aliphatic methylene groups present in lignin
3340–3360	For H-bonded H-O Stretching for chain to chain H-bonding in jute constituents

Additional appearance of transmittance peaks between 1300 cm^{-1} to 1400 cm^{-1} is attributed to C-N stretching vibration in primary and secondary amines due to the incorporation of urea in the formulation. Increase in the band intensity of the peak at 1446 cm^{-1} is ascribed the presence of -N-CH_2 component in the treated jute, supporting reaction-1 to 3 and reaction-6 of reaction scheme 1. The said treatment results in some chemical addition/subtraction and modification of both hemicelluloses and celluloses of jute, which is evident from FTIR spectrum (b), Fig. 1, which shows some decrease in band size at 2920 cm^{-1} and 4000 cm^{-1} band in FTIR spectrum (b), Fig. 1 as compared to spectrum-(a), vide Fig. 1 for bleached jute. Spectrum (b), Fig. 1 also shows a sharp peak in it at 2370 cm^{-1} for N–H vibration of urea present in the jute fabric treated with formulation 1e indicating incorporation of sulfonated-urea on to jute as per reaction scheme 1, discussed earlier above.

J-U-SA—that is, formulation 2f treated jute [Spectrum (c), Fig. 1]—shows some minor changes in functional group pattern by the said treatment when compared with both the spectrum (a) and spectrum-(b), Fig. 1. Thus, In spectrum (c), Fig. 1, for J-U-SA-formulation 2f treated jute, there is a difference in the band intensity at 810 cm^{-1} (indicating the presence of $–SO_3H$) and a duplet arrived at 989 cm^{-1} and 1030 cm^{-1} (due to $–OSO_2$-group of sulfamic acid). There is an increase in band intensity of the transmittance peaks between 1700 cm^{-1} to 1400 cm^{-1}, which is attributed to more acidic degradation of celluloses/hemicelluloses (in more acidic condition there is a partial loss of hemicellulose and cellulose by chain scission) as compared to ammonium sulfamate treatment by formulation 1e. Also the new sharp peak in this spectrum (c), Fig. 1 at 2370 cm^{-1} is also observed for N–H vibration of urea present in the formulation 2f treated jute. No additional -C-N stretching vibration for primary and secondary amines and no increase in the band intensity of the peak at 1446 cm^{-1} which can be ascribed to the presence of -N-CH_2 component in the formulation 1e treated jute, that is, 15 % SA + 12.5 % urea with 3 % $MgCl_2$ catalyst treated jute is found, leading to clear indication that Reaction 8–10 do not occur probably due to low curing temperature (130 °C for 3–4 min) than the degradation temperature of sulfamic acid, which is known to be 205 °C.

Bl Jute- Bleached control Jute; J-U-AS-Formulation 1e; J-U-SA- Formulation 2f; J-Sara-CWF-Formulation 3d; J-U-AS-A-C- Formulation 4

For comparison purpose, the bleached control jute is treated with a commercially available saraflam-CWF, a THPC-based commercial fire-retardant chemical. Thus, FTIR spectra [spectrum (e), Fig. 1] of J-saraflam-CWF, that is, formulation 3d treated jute as compared to the FTIR spectra obtained for the bleached jute (Spectrum –*a*, Fig. 1) shows distinctively some changes in functional group pattern after the said saraflam-CWF treatment, despite showing some minor decrease/increase or weakening/strengthening of some FTIR peaks/bands also. In spectrum (e), Fig. 1, there is an introduction of a new peak at 1646 cm^{-1} (indicating –C-O stretching vibration) and there is an increase in band intensity at 2355 cm^{-1} (due to N–H stretching vibration for associated amines). This fire-retardant treatment results in blocking of C_6 position of -CH_2-OH-group of cellulose due to phosphorylation (from saraflam-CWF), which is evident from FTIR spectrum (e), Fig. 1, as compared to spectrum (a), Fig. 1, which shows some decrease in band size at 3350 cm^{-1}as compared to the corresponding bands in FTIR spectrum (a), Fig. 1 for bleached control jute.

J-U-AS-A-C -formulation 4 treated jute [Spectrum (d), Fig. 1], is a modified formulation of J-U-AS Formulation 1e, additionally admixed with a commercially available acrylic poly-carboxylate binder with polyolefin-based softener for improving the wash durability of the said treatment, where it largely retains similar characteristics in their FTIR spectra as that obtained for the J-U-AS-formulation 1e treated jute in spectrum (b), Fig. 1, thereby only showing some minor decrease/increase or weakening/strengthening of some FTIR peaks/bands. One such minor change is the appearance of a duplet at 2750 cm^{-1} region denoting the presence of polyolefin attributed by $–CH_2–$ asymmetric and symmetric stretching vibration (as an additional polyolefin characteristics for the incorporation of polyethylene-emulsion softener present in ceralube-HD softener along with the arcofix

–NEC acrylic poly-carboxylate binder to improve the wash stability of the said fire-retardant formulation 1e). Finally, from this preliminary study, the superiority of formulation 1e (15 % AS + 12.5 % urea + 3 % $MgCl_2$ catalyst) is considered to be the most acceptable and more particularly modified formulation 1e, that is, formulation 4, is more acceptable for its better wash stability. But still, it needs to optimize the concentration of ammonium sulfamate and urea with the study of their interactive effect for maximizing LOI and minimizing both char length and loss of fabric tenacity, following response surface methodology with user-defined Quadratic Model (UDQM) model of statistical optimization technique.

3.5 Statistical Optimization of the Fire-Retardant Formulation Using Urea and Ammonium Sulfamate Combination by Using UDQM (User-Defined Quadratic Model) Under Response Surface Methodology

Response surface methodology (RSM) has recently emerged as useful mathematical and statistical tools for empirical model building for evaluation of interactive effects amongst process factor, process variables for assessing the optimal process conditions [13]. In statistics, response surface methodology (RSM) explores the relationships between several experimental input variables and the required number of one, two or more response or resultant output variables. The main idea of RSM is to use a sequence of designed experiments to obtain an optimal response. By careful selection of the suitable design of experimental model such as Box and Behnkan or UDQM (user-defined quadratic model), the objective is to generate useful response surface equations to optimize the responsive resultant/output variable by using Response surface methodology (RSM), which is influenced by several independent input process variables.

The design of the experiment is a series of experimental tests, called runs, designed in a planned scientific manner in which equal differences of variables are changed in the input variables in order to identify the assessment of corresponding changes in the resultant output response. As revealed by a series of preliminary experiments, the required average dosages of flame-retardant chemicals and additives/binder, etc., and average conditions of treatment, etc., are to be identified before selecting critical input variables to vary and other variables to be pre-fixed at test run set of experiments of design. Hence, in this part of the work, by use of this response surface methodology (RSM), a suitable statistical optimization model such as UDQM technique of design of experiment has been used for determining the two selective input process variables such as the concentration of Ammonium sulfamate and concentration of urea for optimizing three important resultant output process variables such as limiting oxygen index (LOI) value, char length and loss of fabric Tenacity for corresponding treated fabrics, for the said selective flame retardant formulation along with pre-fixed dosages of catalyst and other additives (e.g. for formulation 1e: in this case, it is concentrations of ammonium sulfamate and urea combination with the pre-fixed amount of $MgCl_2$ catalyst and pre-fixed dosages of acrylic binder and PE-softener as an additive for formulation 4, as discussed above). The concentration of catalyst and additives such as binder and softener has a little role in fire-retardancy performance and similarly the process and curing conditions were kept constant as an increase of the time or temperature causes yellowing and degradation of jute fabric as revealed from the above preliminary study.

Thus, for this statistical optimization process by experiments of design, the new set of the experimental test run for the said selective fire-retardant formulation, a new set of experimental test run are carried out by applying varying concentrations of ammonium sulfamate and urea combination as main fire-retardant chemicals to apply on bleached control jute fabric in accordance with user-defined quadratic model design (UDQM), to obtain the optimized resultant output, that is, maximum limiting oxygen index, that is LOI % (R1), minimum char length in mm (R2) and minimum loss in fabric tenacity in % (R3) as major three response/resultant variables dependent on the said two critical/input process variables such as concentration of urea, and concentration of ammonium sulfamate, for statistical optimizing of formulation 1e (15 % ammonium sulfamate + 12.5 % urea with pre-fixed $MgCl_2$ catalyst of 3 % (1/5th of AS as discussed above), to get the experimental test run parameters by choosing each of the input variables at five different level (–2, –1, 0 + 1, + 2), that is ± 2 levels to perform RSM experiments using Design expert software with other operating (independent) variables pre-fixed [for fixed concentration of catalyst, additives and fixed process and curing conditions (130 °C for 3–4 min) as input treatment variables] for new set of test run for experiments of design model (UDQM) as shown in Table 5.

With the above design of experiments by UDQM at five-level variations, Table 6 represents the values of response variables (R1-LOI, R2-Char Length, R3-Loss in Fabric Tenacity) against the respective independent process variables A (urea) and B (ammonium sulfamate) selected at five levels (–2, –1, 0, + 1 and + 2) as per the UDQM model derived from a specific experiment of design expert software. Relevant results are shown in Table 6 indicating all the resultant response surface variables for

Table 5 UDQM Experimental Design matrix with ranges and levels of the input process variables

Chemicals used	Coded factors	Units	Uncoded values	Coded values
Urea	A	g/l	50.00	– 2
			87.50	– 1
			125.00	0
			162.50	+ 1
			200.00	+ 2
Ammonium sulfamate	B	g/l	50.00	– 2
			100.00	– 1
			150.00	0
			200.00	+ 1
			250.00	+ 2

the corresponding variation in two input process variables A and B as given below. As per the new test run set, the RSM design layout UDQM experimental design technique is used to vary one parameter at a time while keeping the other constant as given in Table 6 and all the experiments are repeated three times and data represented for the response variables (in Results and discussion) are the average value of those three repeat test results in each case.

Thus, Table 6 shows the values of the process variables (A & B) used in the statistical experimental design with user-defined quadratic model (UDQM) run in design expert software and the corresponding resultant response variables (R1, R2, R3) obtained from the test results of those treated jute fabrics under the corresponding experimental set for each case.

These results are then processed for analysis of variance (ANOVA) to obtain the F value and p values (as shown in Tables 8 and 9) for R1-LOI, R2-Char Length, R3-Loss of fabric tenacity.

The corresponding ANOVA tables for three major response variables for LOI (R1) is shown in Table 7, ANOVA table for response variables of char length (R2) is shown in Table 8 and ANOVA table for response variables for loss in fabric tenacity (R3) is shown in Table 9, gives corresponding regression coefficient values of A and B.

Table 6 Values of response variables against the respective independent Process Variable as Per the UDQM Model

Independent process variables		Response variables		
Urea (g/l) (A)	Ammonium sulfamate (g/l) (B)	LOI (%) (R1)	Char length (mm) (R2)	Loss in tenacity (%) (R3)
125 (0)	50 (−2)	27.5	51	16
200 (+2)	50 (−2)	28.3	43	18
50 (−2)	50 (−2)	24.5	56	13
162.5 (+1)	100 (−1)	32.8	36	26
125 (0)	100 (−1)	32	35	28
125 (0)	150 (0)	37.5	15	30
50 (−2)	150 (0)	35.6	22	26
200 (+2)	150 (0)	39.3	9	29
162.5 (+1)	150 (0)	39.2	10	28
87.5 (−1)	150 (0)	35.4	21	27
87.5 (−1)	200 (+1)	35.5	22	30
162.5 (+1)	200 (+1)	39.5	9	33
125 (0)	200 (+1)	38.4	11	32
200 (+2)	250 (+2)	39.5	10	36
125 (0)	250 (+2)	38.6	10	34
50 (−2)	250 (+2)	33.9	19	36

Table 7 ANOVA table for response surface quadratic model for LOI (R1)

Source	Sum of squares	df	Mean square	F value	p-value Prob > F	Remarks
Model	335.61	5	67.12	60.5	< 0.0001	Significant
A-A	43.2	1	43.2	38.94	< 0.0001	
B-B	217.62	1	217.62	196.16	< 0.0001	
AB	1.36	1	1.36	1.22	0.2926	
A^2	0.11	1	0.11	0.095	0.7634	
B^2	64.8	1	64.8	58.41	< 0.0001	
Residual	12.2	11	1.11			
Cor total	347.81	16				

Table 8 ANOVA table for response surface quadratic model for char length (R2)

Source	Sum of squares	df	Mean square	F value	p-value Prob > F	Remarks
Model	3650.88	5	730.18	43.98	< 0.0001	Significant
A-A	282.13	1	282.13	16.99	0.0017	
B-B	2745.63	1	2745.63	165.36	< 0.0001	
AB	0.13	1	0.13	7.97E-03	0.9305	
A^2	4.59	1	4.59	0.28	0.6093	
B^2	591.7	1	591.7	35.64	< 0.0001	
Residual	182.65	11	16.6			
Cor Total	3833.53	16				

Table 9 ANOVA table for response surface quadratic model for loss in tenacity (R3)

Source	Sum of squares	df	Mean square	F value	p-value Prob > F	Remarks
Model	600.43	2	300.22	44.03	< 0.0001	Significant
A-A	10.8	1	10.8	1.58	0.2288	
B-B	589.63	1	589.63	86.48	< 0.0001	
Residual	95.45	14	6.82			
Cor Total	695.88	16				

Thus, to establish the relationship between the variables, analysis of variance (ANOVA) and regression analysis are done. The coefficients generated by analysis of variance (ANOVA) in design expert software, are used to determine the RSM equations—which are shown as Eq. 5 for R1 (LOI), as Eq. 6 for R2 (char length) and as Eq. 7 for R2 (loss of fabric tenacity) for determination of predicted values of particular response variables. Relevant data in corresponding ANOVA tables for all the three response variables (R1, R2 and R3) against A and B as process variables elucidate the lack of fit in terms of p value ($p > 0.05$), which are therefore neglected and used to generate the resultant RSM equations in coded form as expressed below by following equations—as Eqs. 5–7 for R1 (for LOI), R2 (for char length) and R3 (for loss in tenacity), respectively. Putting the value of corresponding regression coefficients, the predicted values of respective resultant variables can be determined/predicted from corresponding RSM equations by following Eqs. 5–7 as given below:

$$R1 = 14.74 + 0.03 * A + 0.19 * B + 7.53 * 10^{-5} * A * B - 3.47 * 10^{-5} * A^2 - 4.83 * 10^{-4} * B^2 \quad (5)$$

$$R2 = 87.05 - 0.03 * A - 0.63 * B + 2.35/100000 * A * B - 2.29/10000 * A * A + 1.46/1000 * B * B \quad (6)$$

$$R3 = 12.35 + 0.02 * A + 0.09 * B \quad (7)$$

It is apparent from the above three ANOVA tables (Tables 7, 8 and 9) that B (ammonium sulfamate) is the most prominent factor followed by A (urea) affecting the three resultant response variables. Both the two factor A and B have synergetic effects on the three response variables (R1-LOI, R2-Char Length, R3-Loss in Fabric Tenacity). Three contour plots for three different response variables (for R1, R2 and R3) are shown in Figs. 2, 3 and 4, respectively.

Thus, the optimal values of A and B as generated by RSM technique by UDQM model experiment of design expert software are found for A (urea) = 87.90 gpl (or 8.79 %) and for B (ammonium sulfamate) = 147.80 gpl (or 14.78 %) for which the resultant optimal values of R1 (LOI), R2 (Char length) and R3 (Loss of tenacity), respectively, are LOI = 35.6 %,Char length = 2.2 cm and Loss in fabric tenacity

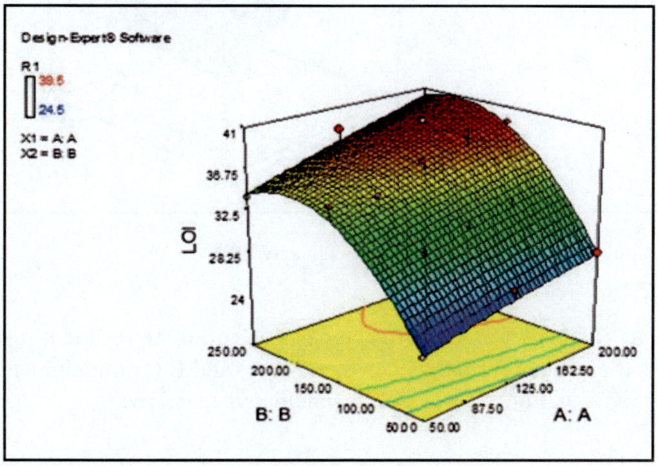

Fig. 2 RSM Contour Plot in respect to LOI (R1)

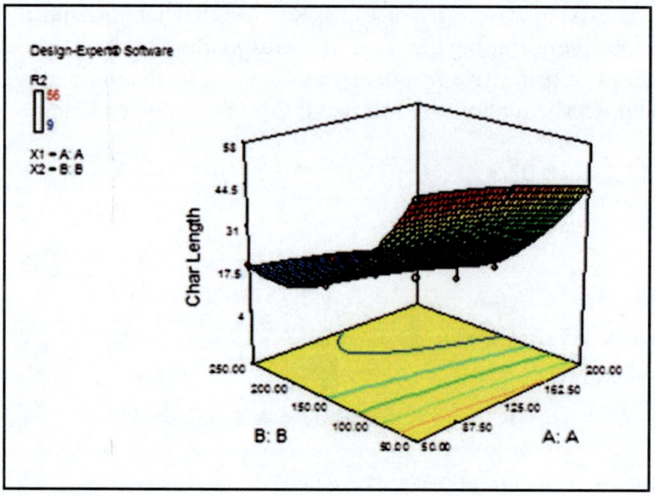

Fig. 3 RSM Contour Plot in respect to Char Length (R2)

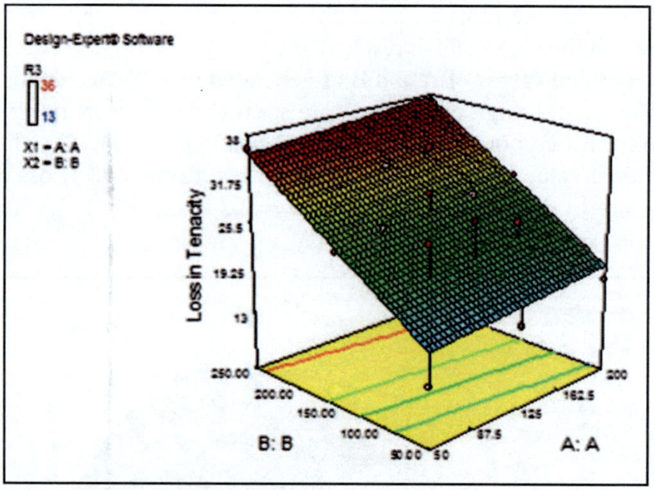

Fig. 4 RSM Contour Plot in respect to Loss in Tenacity (R3)

= 27.39 %, as calculated/predicted values of the said three resultant response variables against the actual optimal values of A (urea) and B (ammonium sulfamate) in respective RSM equations 5–7 as finally calculated as below:

$$
\begin{aligned}
\mathbf{R1} = \text{LOI} &= 14.74 + 0.03 * \text{A} + 0.19 * \text{B} + 7.53 * 10^{-5} * \text{A} * \text{B} \\
&\quad - 3.47 * 10^{-5} * \text{A}^2 - 4.83 * 10^{-4} * \text{B}^2 \\
&= 14.74 + 0.03 \times 87.90 + 0.19 \times 147.8
\end{aligned}
$$

$$+ 7.53 \times 10^{-5}\text{x } 87.9 \times 87.9 - 4.83 \times 10^{-4} 147.8 \times 147.8$$
$$= 14.74 = 2.64 + 28.08 + 0.978 - 0.268$$
$$- 10.55 = \mathbf{35.6}\,\% \quad (8)$$

$$\mathbf{R2} = \text{Char Length} = 87.05 - 0.03 * 1 - 0.63 * \text{B} + 2.35/100000 * \text{A} * \text{B}$$
$$- 2.29/10000 * \text{A} * \text{A} + 1.46/1000 * \text{B} * \text{B}$$
$$= 87.05 - 0.03\text{x}87.90 - 0.63\text{x}147.8 + 2.35/100000\text{x}87.90 \text{ x } 147.8$$
$$- 2.29/10000\text{x}87.90 \text{ x}87.90 + 1.46/1000\text{x}147.8\text{x}147.8$$
$$= 87.05 - 2.64 - 93.114 + 0.305 - 1.751 + 31.89 = 21.74 \text{ mm}$$
$$= \mathbf{2.17\ cm.} = \mathbf{2.2cm} \quad (9)$$

$$\mathbf{R3} = \text{Loss of Fabric Tenacity} = 12.35 + 0.02 * \text{A} + 0.09 * \text{B}$$
$$= 12.35 + 0.02\text{x}87.90 + 0.09\text{x}147.8 = 12.35 + 1.742 + 13.30 = 27.39\% \quad (10)$$

The contour plots of the effects of independent process variables individually or in combination on the resultant values of LOI (Fig. 2), Char length (Fig. 3) and loss of fabric tenacity (Fig. 4) are given here for understanding both the effects of one single process variables and also for interactive combined effect for different combinations (AA, AB, BB), that is interdependent factors also. The regression coefficient obtained from optimization by design expert software following UDQM model used for RSM optimization technique either have positive or negative values and so have positive or negative effects on the experimental results (for the properties selected and studied). For a process variable to have a significant effect, the coefficients must be greater than twice the standard error. However, non-significant coefficient also has some inputs and is not useless or not to be dismissed, as there may be some small effect, but is important too. Thus from these three plots of contour diagram, it can be said that the effects of independent process variables or their combined interrogative effects on R1 (LOI), R2 (char length) and R3 (Loss of Tenacity), may be now interpreted easily and can be explained/understood easily from the corresponding plots (Figs. 2, 3 and 4) and corresponding RSM Equations (RSM equations 5–7).

RSM contour plots in Fig. 2 shows that variable A (i.e., Urea) has a less positive effect on the increase of LOI value; showing an increase in LOI value with an increase in the concentration of AS (B), which, however, is not that significant after 87.5 gpl, while the effect of variable A is not that much after increasing it to 147.8 gpl. Thus, corresponding RSM equation 5 generated for predicting the value for R1 (LOI) for specific dosages of A (Urea) and B (AS) indicate that the positive values of both (A) and (B) has a positive effect, while negative values of the coefficients for both A^2 and B^2 terms, meant that further increase of the dosages for both the chemicals (urea + AS) will reduce or have no effect on increasing LOI value, rather may reduce or remain same. The positive and higher value of coefficients for the interactive effect between AS (B) and Urea (A) meant the effect is positive and significantly higher, that is there will be a much higher effect obtained for using a suitable combination of urea and AS than the use of AS individually and use of Urea individually. For an increase in LOI value, this combination of urea and AS has a synergistic effect. So, it may be mentioned here that both the chemicals are equally required to get the optimum value of LOI and other response variables.

RSM contour plot in Fig. 3 shows that char length (one of the major second important criteria for the fire-retardant performance of textiles) is reduced by increasing both the process variables (A) and (B) up to a certain dosages and then further increases above a certain dosage level of these process variables. However, the effect of variable (A), that is urea is found to be lower than the effect of variable (B), that is, ammonium sulfamate.

Corresponding RSM equation 6 generated for predicting values of R2 (char length) for specific dosages of urea (A) and ammonium sulfamate (B) indicate the negative value of the coefficient for A and B and A^2 term, that is, both increase of A & B cause a good reduction in char length but very high increase in A (Urea) did not show much reduction in R2 (char length). Hence, it can be said that ammonium sulfamate is much more important and is essentially required to reduce char length along with urea. So, the determination of optimal values for the combination of ammonium sulfamate and urea is essential in this treatment formulation.

RSM contour plot in Fig. 4 shows that loss in fabric tenacity of treated jute fabric is increased with an increase in input variable B (ammonium sulfamate) to a large extent than the effect of variable A, that is, urea. Increase in concentration of urea has a minimum or very less effect on loss of fabric tenacity. Ammonium sulfamate partially produces sulfamic acid and also use of $MgCl_2$ acidic catalyst causes an acidic environment particularly during heating (at curing stage) and hence, there is probable acidic degradation of jute cellulose/hemicelluloses chains, causing the said loss in fabric tenacity, where acidic degradative action of cellulose chains is enhanced due to use of pre-fixed small dosages of $MgCl_2$ catalyst used in all such fire-retardant formulations used.

Corresponding RSM equation 7, generated for predicting resultant values of R3 (loss in tenacity) for specific dosages of Urea (A) and ammonium sulfamate (B) indicate the positive value of coefficient, for both variables A and B showing coefficient of variable—B is higher than coefficient—A, that is it meant the effect of increase of Ammonium sulfamate is much higher than the said effect of increase of variable A, that is Urea. Hence, the dosages of AS is to be restricted to avoid higher loss in fabric tenacity. So there is no question of using Urea and AS in a higher dosages as understood from values of A^2 and B^2 terms.

Finally, the numerical optimization program was run in the design expert software of UDQM model to evaluate the optimal parametric values which are correspondingly represented in the three respective contour curves in plots given in Figs. 2, 3 and 4 presenting the above-stated results mentioned above for R1 (LOI), R2 (Char length) and R3 (Loss of fabric tenacity), respectively.

4 Conclusions

So, it may be finally concluded from the above study that optimum dosages of ammonium sulfamate obtained by this statistical optimization technique is 147.8 gpl or 14.8% and the same for Urea is 87.90 gpl or 8.8 %, which are thus much closer to the values of sem-optimal dosages of these two chemicals as earlier obtained from preliminary experiments reported here before statistical optimization for formulation 1e using 10 % urea and 15 % ammonium sulfamate combination (Formulation 1b) in presence of 3 % $MgCl_2$ catalyst (1/5th of weight of ammonium sulfamate) for which LOI is 34, char length is 2.5 cm and loss in tenacity value is 26 %, though formulation 1e shows LOI-value is 38, char length is 2 cm and loss of fabric tenacity is 28 %.

Finally, considering the interactive effect of these two chemicals the finally statistically optimized results for determining optimum dosages of these two chemicals are finally arrived as treatment with 8.8 % of urea and 14.8 % of ammonium sulfamate in presence of 3 % $MgCl_2$ as catalyst (1/5th amount of weight of ammonium sulfamate) with the addition of pre-fixed dosages of 50 gpl or 5 % arkofix –NEC acrylic–poly-carboxylate binder and 30gpl or 3 % ceralube–HD (polyolefin-based emulsion softener) by Pad–Dry (at 100 °C for 10 min) and Curing (at 130 °C for 3–4 min) technique followed by normal washing and drying, for obtaining somewhat little lower fabric tenacity loss of 27.39 % (instead of 28 % as is found in formulation 1e in preliminary experiment), LOI-35.6 and char length-2.2 cm, with somewhat reduced concentration of urea than that used in formulation 1e (thus saving the cost of excess amount urea in this optimized recipe, than that used for formulation1e, making more industry acceptable recipe).

Acknowledgements The authors are grateful to Govt of India, Ministry of Textiles (MOT), Delhi and Office of the Jute Commisioner, Govt of India, Kolkata for required financial support in the form of R & D Project grant-in–aid for the project entitled "Standradization of Fire Retardant Finish for Jute Fabrics" provided to this Department of University of Calcutta (DJFT, IJT, CU) vide Ref Sanction letter no-nil dt 08-03-2016 from MOT, GOI. Moreover, Prof AKS as one of the author is especially thankful to CU-TEQIP Cell for providing the required travel and other logistic support to present this paper at FTC-2020 International conference at IIT-Delhi on 7–9 February-2020.

References

1. Varma IK, Anantha Krishnan SR, Krishnamoorthy S (1988) Effect of chemical treatment on thermal behaviour of jute fibers. Text Res J 58(9):486–494
2. Basak RK, Saha SG, Sarkar AK, Saha M, Das NN, Mukherjee AK (1993) Thermal properties of jute constituents and flame retardant jute fabrics. Text Res J 63(11):658–666
3. Basak RK, Saha SG, Sarkar AK, Saha M, Das NN, Mukherjee AK (1993) Thermal properties of jute constituents and flame retardant jute fabrics. Text Res J 63:658-664
4. Banerjee SK, Day A, Ray PK (1986) Fireproofing jute. Text. Res J 56(5):338–339
5. Samanta A K : Chemical finishing of jute and jute blended textiles: a review. Colourage 42:37–42

6. Samanta AK, Bagchi A, Biswas SK, Bhattacharjee (Roy) R (2011) Semi durable –fire retardant finishing of jute fabric and its thermal behaviour. J Inst Eng (India)-Text Eng 91:18–26
7. Samanta AK, Bagchi A, Biswas SK (2011) Fire retardant finishing of jute fabric and its thermal behavior using phosphorous and nitrogen based compound. J Polym Mater 28(2):151–171
8. Horrocks AR (1986) Flame-retardant finishing of textiles. Rev Prog Coloration 16:62–101
9. Horrocks AR (1996) Developments in flame retardants for heat and fire resistant textiles-the role of char formation and intumescence. Polym Degrad Stab 54:143–154
10. Weil ED, Levchick SV (2008) Flame retardants in commercial use or development for textiles. J Fire Sci 26(3):243–281
11. Ramachandran T, Vellingiri K, Kannan MSS (2005) A comparative study of durable flame retardant finish on cotton fabrics. J Inst Eng (India) Text Eng 85:29–32
12. Zhou X, Chen K, Yi H (2014) Synthesis and application of a formaldehyde-free flame retardant for bamboo viscose fabric. Text Res J 84(14):1515–1527
13. Thilagavathi T, Geetha D (2013) Low-temperature hydrothermal synthesis and characterization of ZnO nanoparticles. Indian J Phys 87:747–750
14. Samanta AK, Bhattacharyay R, Jose S, Basu G, Chowdhury R (2017) Fire retardant finish of jute fabric with nano zinc oxide. Cellulose 24:1143–1157
15. Xia L, Wang H-L, Shentu B-Q, Weng Z-X (2010) Ammonium sulfamate flame retarded polyamide 6: morphology and thermal degradation. Chin J Polym Sci 28:753–759
16. Mostashari SM, Mostashari SZ (2008) Combustion pathway of cotton fabrics treated by ammonium sulfate as a flame-retardant studied by TG. J Therm Anal Calorim 91:437–441
17. Mostashari SM, Mostashar SZ (2008) Combustion pathway of cotton fabrics treated by ammonium sulfate as a flame-retardant studied by TG. J Therm Anal Calorim 91(2):437–441
18. Bajaj P (1992) Fire-retardant materials. Bull Mater Sci 15(1):67–76
19. Tang W, Zhang S, Sun J, Xiaoyu Gu (2016) Flame retardancy and thermal stability of polypropylene composite containing ammonium sulfamate intercalated kaolinite. Ind Eng Chem Res 55(28):7669–7678
20. Lewin M, Zhang J, Pearce E, Gilman J (2007) Flammability of polyamide 6 using the sulfamate system and organo-layered silicate. J Polm Adv Tech 18(9):737–745
21. Lewin M (1997) Flame retarding of polymers with sulfamates. Sulfation of cotton and wool. J Fire Sci 15(4):263–276
22. ISI (BIS) (1982) Handbook of textile testing, SP: 15–1981. Bureau of Indian standard, New Delhi
23. ASTM-D-2863–77 (1994) Annual Book of ASTM Standards. American Society for Testing and Materials, Philadelphia, Pennsylvania, USA
24. ASTM D6413/D6413M-13b (1994) Annual Book of ASTM Standards. American Society for Testing and Materials, Philadelphia, Pennsylvania, USA
25. Samanta AK, Bhattacharyay (Roy) R, Bagchi A,Chowdhuri R (2019) Statistical optimisation of nano zinc oxide based fire protective finish formulation on jute fabric. In: Majumdar A, Gupta D, Gupta S (eds) Functional textiles and clothing. Springer, Singapore, Part – VIII (Textile Chemical Processing), pp 167–192

Application of Protective Finishes on Denim and Analysis of Its Multifunctional Performances

Kundlata Mishra and Ela Dedhia

1 Introduction

Protective clothing (PC) includes all clothing and equipment worn in place of regular work clothing to protect the workers [1]. Though protective clothing is used as the last measure for workplace hazards control and mostly where managerial and engineering control measures are already in place, it surely sets up a barrier against any hazard, thus enhancing the safety of the people working under hazardous conditions. Each potential hazard has different problem areas and requires specific solution in the form of PC [2].

In India, standard protective clothing is not available easily or it is expensive. Additionally, there is a distinct lack of safety culture in India, and in spite of so many process industries the concept of protective clothing is still at an infancy stage. In the majority of the units/industries workers are wearing the available work clothes or the protective suit that is uncomfortable and+ hardly provides any protection against harmful chemicals [3]. Though protective clothing regulations are strictly imposed in the developed nation, India is not having a serious check on such kind of set rules and standards in all industries in spite of Factory Act 1986 and various other Indian standards of industrial work wear.

"Denim is one of the world's oldest fabric, yet it remains eternally young", declared a writer in 1969 for the American Fabrics magazine. Lynn Downey in his 2014 article, "A Short History of Denim" mentions that the fabric is present since the seventeenth century and that there were different terms used for denim [4]. In France, denim being the English term was called "Serge de Nimes" a serge fabric from the

K. Mishra (✉)
Fashion Design Department, National Institute of Fashion Technology, Mumbai, India
e-mail: kundlata.mishra@nift.ac.in

E. Dedhia
Department of Clothing & Textiles, Nirmala Niketan College of Home Science, University of Mumbai, Mumbai, India

A. Majumdar et al. (eds.), *Functional Textiles and Clothing 2020*,
https://doi.org/10.1007/978-981-15-9376-5_10

town of Nimes before seventeenth century as said by most of the reference books. This fabric was partly composed of wool. Denim was reserved for work clothes due to its durability and comfort.

Denim is strong and serviceable, has strong warp direction, and is considered as the most important fabric of work-clothing group as said by John Hoye in "Staple cotton fabrics" textbook. Hence, denim is used for protective clothing, which gives comfort to the wearer with its strength to hold tension, abrasion, and other physical properties [5].

For electrical workers, safety on the job begins when they get dressed at home. Street clothes are actually the first layer of protection against the heat of an arc flash. In addition to any required PPE, an electrical worker is expected to wear natural fiber (usually cotton) street clothing and underwear. Synthetic fibers such as nylon and polyester are not allowed because they may melt and adhere to the skin, resulting in secondary burns. Proper street clothes can be required as a condition of employment, in the same way that a suit is the dress code in some offices [6].

This study aims to blend the idea of exuberant research that has gone into the field of protective clothing right from material selection to modeling with emphasis on creating a fashion statement with safety. The idea is to infuse style in such a way that every worker of the industrial plants looks forward to wearing PCs, which they are proud of and comfortable to carry, create no disparity among the workers, and yet keep them safe [7].

There is a need to develop protective clothing that is lightweight, comfortable, economical, and acceptable to workers [8]. The objective of this study is to provide new comfortable design solutions for industrial workers. During the research, it was found that the main reason for noncompliance, according to a third of workers was feeling of invulnerability [9]. The workers in industries performing welding, spray painting, and other industrial fabrication jobs in the workshops were exposed to extreme heat in petrochemical industries and rarely use PPEs and other protective measures that can minimize health risks. The other risk factors observed were falling and flying objects, sharp edges, dust, vapors, and fumes/odors from chemicals and heat [10]. The purpose of the study is to serve the market for PC with high-quality resolution; hence the study focused on 100% cotton. The finished fabric was found to be effective against heat and flame, anti-stain, and aesthetically appropriate. For the protective material, 100 % cotton was selected for its comfort and aesthetic value. Job profiles were identified and various combinations of finishes were tested and the most appropriate ones were selected and applied. Denim wash was done with enzyme to make the process ecofriendly.

Table 1 Details about denim fabric specifications

Fabric specification	Content type	100 % Cotton	
Weave		3/1 Twill	
Ounce (three types for every finish)	9 oz	11 oz	13 oz
No. of ends and picks per inch	66 × 50	59 × 45	55 × 42
Warp twist count	28 s	32 s	34 s
Weft twist count	22 s	20 s	16 s

2 Materials and Methods

2.1 Materials

2.1.1 Fabrics

100 % cotton denim fabrics with the following specifications were used (Table 1).

For all the experiments, reactive and/or indigo dyed, de-sized, enzyme/bio-washed, softened (silicon softener) fabrics were used.

2.1.2 Chemicals

Commercially available anti-microbial finishes that are polymer of silane quaternary ammonium compound were selected, namely Sanigard 7500 (L.N. Chemical), Fabsheild AEM 5700, and Microban PB 80150-050 (Rossari Biotech Limited). Commercially available flame-retardant finishes such as Glogard-CP-New with Fixer ALN and Modfinish HDNI (L.N. Chemical) were selected and used in a proportion as per the fabric weights.

Quest-AQ (L.N. Chemical) and Inovasof SRW-N, Roswik 14-H, Rosil GETZ (Rossari Biotech Limited) were the two types of moisture management chemicals selected. Aquarepel XC-830 with Aquarepel XAN and Aquarepel XC-630 (L.N. Chemical) were two stain release or soil release finishes selected.

2.1.3 Preparation of Fabrics

The dark indigo and reactive dyed denim was washed with different chemicals to make the fabric de-sized, less lint producing, soft, and comfortable for the wearer.

The process included the following steps and chemicals:

De-sizing (Table 2):

Enzyme/Bio-washing (Table 3):

Softening (Table 4):

Table 2 Recipe for de-sizing

Content	Weight (g/l)
Anti-back stain	30 g/l
Wetting agent	20 g/l
De-sizer	50 g/l
Lubricant	70 g/l

Table 3 Recipe for enzyme/bio-washing

Content	Weight (g/l)
Anti-back stain	30 g/l
Liquid enzyme	50 g/l
Lubricant	35 g/l

Table 4 Recipe for softening

Content	Weight (g/l)
Silicon softener	50 g/l

At room temperature for 5 min

Hydro-extraction:

Hydro-extraction is carried out in order to remove excess solution and water on the fabric or garment.

Tumble Drying:

This is done to dry the fabric or garment immediately. After the pre-treatment, coding of samples, finish application, and further evaluations were conducted.

2.1.4 Sample Codes

The samples were coded to preliminary work 1, preliminary work 2, Job Profile 1, Job Profile 2, and Job Profile 3 as A, B, C, D, and E, respectively. According to finishes selected, their combinations are coded as 100, 200, 300, and so on; also, GSM 9, 11, and 13 oz as 10, 20, and 30, respectively. In addition, control sample was denoted as CS in each preliminary work and Job Profile. Total number of samples tested was 87 samples (Tables 5, 6, 7, 8, and 9).

2.1.5 Preliminary Work

The preliminary experimental work was carried out to get an insight into the various commercially available anti-microbial finishes, their recommended recipes, also their

Table 5 Sample code A and application of three types of anti-microbial finish on reactive dyed fabric

For preliminary work 1 (coded as "A") where application of three types of anti-microbial finish on three types of denim weight that are reactive dyed fabric was accomplished

Type of finish	Sample codes	9 oz	11 oz	13 oz
-	CS A	010	020	030
1	A	110	120	130
2	A	210	220	230
3	A	310	320	330

Table 6 Sample code B and application of three types of anti-microbial finish on indigo dyed fabric

For preliminary work 2 (coded as "B") where application of three anti-microbial finish was accomplished on indigo dyed fabric

Type of finish	Sample codes	9 oz	11 oz	13 oz
–	CS B	010	020	030
1	B	110	120	130
2	B	210	220	230
3	B	310	320	330

Table 7 Sample code C and application of three types of anti-microbial finish, one flame retardant, and two moisture management finishes

Application of combination finishes for petrochemical industries (coded as "C"). It included three anti-microbial, one flame retardant, and two moisture management finishes

Type of finish	Sample codes	9 oz	11 oz	13 oz
–	CS A	010	020	030
–	CS B	010	020	030
1	C	110	120	130
2	C	210	220	230
3	C	310	320	330
4	C	410	420	430
5	C	510	520	530
6	C	610	620	630

effect on the reactive and indigo dyed fabric and to handle the sophisticated instruments required for measuring various properties; importantly, the microbe-resisting properties.

All the finishes were applied in the same concentration in order to determine the optimum concentration. The samples were coded to preliminary work 1, preliminary work 2, Job Profile 1, Job Profile 2, and Job Profile 3. According to finishes selected GSM as 9, 11, and 13 oz, respectively. Total samples tested and number of samples

Table 8 Sample code D and application of three types of anti-microbial finish and one flame-retardant

Application of combination for welders (coded as "D"). It included three anti-microbial and one flame-retardant finishes				
Type of finish	Sample codes	9 oz	11 oz	13 oz
–	CS A	010	020	030
–	CS B	010	020	030
1	D	110	120	130
2	D	210	220	230
3	D	310	320	330

Table 9 Sample code C and application of three types of anti-microbial finish and two stain release

Application of combination finishes for spray painters and lathe machine workers, industries that relate to these types of work profiles including three anti-microbial and two stain releases				
Type of finish	Sample codes	9 oz	11 oz	13 oz
-	CS A	010	020	030
-	CS B	010	020	030
1	E	110	120	130
2	E	210	220	230
3	E	310	320	330
4	E	410	420	430
5	E	510	520	530
6	E	610	620	630

was 87 samples. The finishes were applied on denim 100 % cotton 3/1 twill weave on both reactive and indigo dyed fabrics using standard recipes and recommended concentrations.

2.1.6 Evaluation of Effect of Concentrations of Antimicrobial Finish on Two Types of Denim

The finished fabrics were evaluated for shrinkage, stiffness or bending length, crease recovery, air permeability, thermal conductivity, water vapor permeability, tensile strength, tear strength, abrasion test, durability to laundry, and anti-microbial properties. On the basis of the results of the preliminary work type of fabric dyed, indigo dyed fabric was shortlisted. Exhaust method, drying temperature and time, pH of bath and procedure were used for applying the finishes [11]. The concentrations were kept constant for better comparison for job profiles in the research. The concentrations varied according to the GSM.

The finished fabrics were evaluated for the following properties:

2.2 Determination of Anti-Microbial

The test method AATCC 100 was used for testing the growth of microorganisms. The microorganisms were grown in liquid culture and this concentration of test microorganism was standardized. The microbial culture was diluted in a sterile nutritive solution. Control and test fabric swatches were inoculated in this liquid. The inoculation was performed such that the microbial suspension touches the fabric. Bacteria levels on both control and test fabric were determined at "time zero", that is "B" by elution in a large volume of neutralizing broth, followed by dilution and plating. A control was run to verify that the neutralization/elution method effectively neutralized the anti-microbial agent in fabric. Additional inoculated control and test fabric were allowed to incubate, undisturbed in sealed jars for 24 h, that is, "A". After incubation, microbial concentrations were determined. *Note:* For AATCC 100, percentage reduction: R = 100 (B – A/B), where "B" and "A" are the number of bacteria per sample observed before and after 24 h of incubation, respectively.

2.3 Determination of Weight Add-On

The weight of the samples was taken before and after the application of finish. The increase in weight gave the percentage add-on in weight. The concentration of anti-microbial finish was increased with the increase in weight of fabric, as there is more absorption capacity.

2.4 Determination of Percentage Shrinkage

The standard ISO 6330:2012 specifies domestic washing and drying procedures for textile testing. The procedure was applicable to textile fabrics, garments, or the textile articles, which are subjected to appropriate combinations of domestic washing and drying procedures.

2.5 Determination of Fabric Stiffness

The bending length was measured on Shirley's stiffness tester, in accordance with ASTM D1388-96 test method.

2.6 *Determination of Crease Recovery*

The crease behavior was as per IS 4681–1981 test method. Five tests of each sample in warp and weft direction were taken and an average value of crease recovery angle was calculated.

2.7 *Determination of Durability of Finishes to Laundry*

The durability of finishes to laundry was determined by subjecting the finish-applied samples to laundry cycles as per ISO 6330:1984 standard. After laundry the samples were evaluated for microbe reaction and/or penetration as per AATCC 100.

2.8 *Determination of Transmission*

Air permeability was determined in accordance with test method ASTM D-737-96. The thermal conductivity test was conducted for the comparison with results obtained using the standard unit and test method ASTM D-1518. Standard test method used for water vapor transmission of materials was ASTM E-96 Procedure B.

2.9 *Determination of Mechanical Properties*

The method IS 1969-1, mainly applicable to woven textile fabrics, was used to determine the tensile strength. The tear strength was measured as per ASTM D-412 standard test method and Elmendorf tear tester. The test method for abrasion ISO 12947–1:1998 pertains to determination of the abrasion resistance of fabrics by the Martindale method.

3 Results and Discussions

The preliminary experimental work was carried out to get an insight into the various commercially available anti-microbial finishes, their recommended recipes, also effect on the reactive and indigo dyed fabric.

The anti-microbial finish was applied in 1:10 MLR exhaust method and with 2 % prepared solution. The same concentration, as given in Table 10, is used for every three types of finishes (Tables 11 and 12).

Table 10 Details about the concentration in recipe for anti-microbial finish

Anti-microbial finishes	Concentration (% o.w.f.)		
Fabric weight	9 oz	11 oz	13 oz
Sanigard-7500	0.5	1	2
Microban PB 80,150–050	0.5	1	2
Fabshield AEM 7500	0.5	1	2

Table 11 Effect of application of anti-microbial finish on reactive dyed sample

Treatment	Anti-microbial (%)		
Standard	AATCC 100		
GSM (Oz)	9 oz	11 oz	13 oz
CS-A	0	0	0
A 1	87	88	88.3
A 2	77.5	88	88
A 3	87.8	88	88.5

Note: For AATCC 100 percentage reduction: R = 100 (B – A/B)

Table 12 Effect of application of anti-microbial finish on indigo dyed sample

Treatment	Anti-microbial (%)		
Standard	AATCC 100		
GSM (oz)	9 oz	11 oz	13 oz
CS-B	99.99	99.99	99.99
B 1	99.99	99.99	99.99
B 2	99.99	99.99	99.99
B 3	99.99	99.99	99.99

Note. For AATCC 100 percentage reduction: R = 100 (B – A/B)

3.1 Job Profile 1: Petrochemical Industries

During the course of work, it was found that the use of indigo dyed fabric would make it essential for its certain ecological reasons. The chemical composition of flame-retardant finish is organic phosphorous compound, that is, fiber reactive/compatible to cotton. The appearance of the chemical is colorless. The active component in the viscous liquid is 80 % and can be applied in spray or pad-dry-cure method giving similar results. Application of flame-retardant finish was then followed with a polymeric hydrophile compound, that is, moisture management finish. This would make the fabric more moisture absorbent. The third chemical application was done to gain moisture management that may be lost due to hydrophobic flame-retardant

Table 13 Concentration in recipe for anti-microbial finishes

Anti-microbial finishes	Concentration (% o.w.f.)		
Fabric weight	9 oz	11 oz	13 oz
Sanigard-7500	0.5	1	2
Microban PB 80150-050	0.5	1	2
Fabshield AEM 7500	0.5	1	2

Table 14 Recipe for flame-retardant finish

Flame Retardant	
Glogard-CP-New	300–400 g/l
Fixer ALN	50–80 g/l
Modfinish HDNI	30–40 g/l
Phosphoric acid	20–25 g/l
pH	6–6.5

Denim was then subjected to flame-retardant finish 2 dip-2 nip
Pad > Dry (120° for 5 min) > Cure (160° for 10 min)

finish. Flame retardant is tested according to Indian Standard IS 15025 and included in OSHA, also for industrial act.

The anti-microbial finish was applied in 1:10 MLR exhaust method and with 2 % prepared solution. The same concentration as given in Table 13 is used for every three types of finishes (Tables 14, 15, 16 and 17).

Table 15 Recipe for moisture management finishes type 1

GSM	9 oz	11 oz	13 oz
Moisture management (Type 1):			
Inovasof SRW-N	1 % o.w.f	2 % o.w.f	3 % o.w.f
Rowik 14-H	3 % o.w.f	5 % o.w.f	7 % o.w.f
Rosil GETZ	3 % o.w.f	5 % o.w.f	7 % o.w.f

Denim was applied with moisture management type 1 in 2 dip-2 nip.
Pad > Dry (120° for 5 min) > Cure (130° for 10 min)

Table 16 Recipe for moisture management finishes type 2

GSM (oz)	9 oz	11 oz	13 oz
Quest-AQ-15	10 g/l	15 g/l	20 g/l
pH	5–7		

Denim was applied with moisture management type 2 in 2 dip-2 nip
Pad > Dry (room temperature)

Table 17 Effect of application of anti-microbial, flame retardant, and moisture management finish on indigo dyed sample

Treatment	Anti-microbial (%)			Flame retardant			Water vapor permeability ($g/m^2/day$)		
Standard	AATCC 100			ISO 15,025			ASTM E-96, Part B		
GSM (oz)	9 oz	11 oz	13 oz	9 oz	11 oz	13 oz	9 oz	11 oz	13 oz
CS- B	99.99	99.99	99.99	Poor	Poor	Poor	820.1	823.8	831
C 1	99.99	99.99	99.99	Excellent	Excellent	Excellent	752.3	753.7	789.2
C 2	99.99	99.99	99.99	Excellent	Excellent	Excellent	801.3	808.6	808.9
C 3	99.99	99.99	99.99	Excellent	Excellent	Excellent	802.6	808.5	810.2
C 4	99.99	99.99	99.99	Excellent	Excellent	Excellent	811.5	813.1	820.6
C 5	99.99	99.99	99.99	Excellent	Excellent	Excellent	805.9	808.7	811.4
C 6	99.99	99.99	99.99	Excellent	Excellent	Excellent	791.3	796.8	799.1

Note. For AATCC 100 percentage reduction: R = 100 (B – A/B). Flame retardant sample ignition time was 10 s

Table 18 Concentration in recipe for anti-microbial finishes

Anti-microbial finishes	Concentration (% o.w.f.)		
Fabric weight	9 oz	11 oz	13 oz
Sanigard-7500	0.5	1	2
Microban PB 80,150–050	0.5	1	2
Fabshield AEM 7500	0.5	1	2

The anti-microbial finish was applied in 1:10 MLR exhaust method and with 2 % prepared solution. The same concentration as given in Table 18 is used for every three types of finishes

Table 19 Recipe for flame-retardant finish

Flame retardant	
Glogard-CP-New	300–400 g/l
Fixer ALN	50–80 g/l
Modfinish HDNI	30–40 g/l
Phosphoric acid	20–25 g/l
pH	6–6.5

The denim was then subjected to flame retardant finish 2 dip-2 nip Pad > Dry (120° for 5 min) > Cure (160° for 10 min)

3.2 *Job Profile 2: Welders*

A comparative assessment of unfinished denim and commercially available flame retardant and moisture management combination finish applied denim was conducted. The evaluated results of water vapor permeability were found to be satisfactory. Hence, experimentation on selection of best finish was being carried out and it was found that moisture management finishes were needed to be withdrawn due to the observed results. In view of this, application of anti-microbial and flame-retardant finishes was taken up and evaluated. The recipe followed for anti-microbial and flame-retardant finishes was the same as that of Job Profile 1 (Tables 19 and 20).

3.3 *Job Profile 3: Spray Paint*

Chemical combination of anti-microbial and stain/oil release finish was applied to impart these properties to denim fabric that could then be used for protective clothing. Conventionally, Fixer ALN was incorporated in the recipe to improve the durability of the finish to laundry. Similar research for fluorocarbon finish with DMDHEU

Table 20 Effect of application of anti-microbial and flame-retardant finish on indigo dyed sample in Job Profile 2 (Welders)

Treatment	Anti-microbial (%)			Flame retardant		
Standard	AATCC 100			ISO 15,025		
GSM (oz)	9 oz	11 oz	13 oz	9 oz	11 oz	13 oz
CS-B	99.99	99.99	99.99	Poor	Poor	Poor
D 1	99.99	99.99	99.99	Excellent	Excellent	Excellent
D 2	99.99	99.99	99.99	Excellent	Excellent	Excellent
D 3	99.99	99.99	99.99	Excellent	Excellent	Excellent

Note: For AATCC 100 percentage reduction: R = 100 (B − A/B); Flame retardant sample ignition time was 10 s

Table 21 Recipe for stain/oil repellent finish (type 1)

Stain/oil repellent finishes (Type 1)	
Aquarepel XC-830	400–500 g/l
Fixer ALN	50–80 g/l
Aquarepel XAN	60–70 g/l
pH	5–5.5

Table 22 Recipe for stain/oil repellent finish (type 2)

Stain/oil repellent finishes (Type 2)	
Aquarepel XC-630	40–100 g/l
Fixer ALN	50–80 g/l
pH	5–5.5

The denim was then subjected to stain repellent finish 2 dip-2 nip Pad > Dry (110° for 5 min) > Cure (160° for 10 min)

(dimethyloldihydroxy ethylene urea) based resin was used to impart durability of fluorocarbon finish, and also to impart anti-crease property (Audenuert, 1999). The recipe followed for anti-microbial finishes was the same as that of Job Profile 1 (Tables 21 and 22).

3.4 Evaluation of Effect of Combination of Finishes

The finished applied denim samples were evaluated for various properties (water vapor permeability, water absorption test, weight add-on, percentage shrinkage, stiffness, crease recovery, durability to laundry, air permeability, thermal conductivity, tensile strength, tear strength, and abrasion tests). The denim samples were also tested for the combination finish properties as per the preliminary test and job profiles such

Table 23 Effect of application of anti-microbial and stain release finish on indigo dyed sample in Job Profile 3 (spray paint and lathe machine)

Treatment	Anti-microbial (%)			Stain release		
Standard	AATCC 100			AATCC 130		
GSM (oz)	9 oz	11 oz	13 oz	9 oz	11 oz	13 oz
CS-B	99.99	99.99	99.99	3	3	3
E 1	99.99	99.99	99.99	4	3–4	4
E 2	99.99	99.99	99.99	4	4	4
E 3	99.99	99.99	99.99	4	4	3–4
E 4	99.99	99.99	99.99	4	4	3–4
E 5	99.99	99.99	99.99	4	4	4
E 6	99.99	99.15	99.99	4	4	4

Note Grades are given from 1 to 5, in which, grade 5 shows the best release of stain and grade 1 as poor

as anti-microbial, flame retardant properties, and stain repellent. The results were tabulated and interpreted to arrive at the optimum combination (Table 23).

Efficacy of the Selected Finishes

It was evident that the combination.

1. Anti-microbial [Sanigard 7500 (L.N. Chemical), Fabsheild AEM 5700 and Microban PB 80,150–050 (Rossari Biotech Limited)] and flame retardant (Organo phosphorous compound, i.e., Glogard CP new).
2. Anti-microbial and stain release [Aquarepel XC-630 (L.N. Chemical)]

Offer a good option for a durable anti-microbial and flame retardant; anti-microbial and stain, oil release finish applicable on 100 % cotton denim fabrics.

- Addition of Fixer ALN to the finishes improves its durability to laundry. The chemicals used do not adversely affect the other properties.
- The garments were found to be effective against microbe penetration, flame or fire spark penetration, and stain or oil releasing, as shown by the laboratory results.
- Special design feature in construction of garment was incorporated to ensure adequate protection against hazards and better carrying capacity for tools and other things.
- The use of protective clothing is essential for the protection of workers in process and original equipment manufacturer (OEM) industries. Use of protective clothing developed is also recommended for workers in any other chemical industries.

Effect of Type of Finishes on Various Properties

Weight Add-On

After application of anti-microbial finish denim showed a weight add-on of about 0.6–1.1 %. Hence an increase of weight by 0.6–0.8, that is 9 oz and 11 oz denim was suitable for wearers comfort in protective clothing.

Shrinkage Test

The percentage shrinkage test conducted resulted in lowering of shrinkage as compared to control sample. Maximum shrinkage was observed in 13 oz fabric where control sample (-) 2.9 % and finish applied sample in the range of (-) 2.5 % to (-) 2.7 %. A gradual increase in shrinkage percentage was observed from 9 to 13 oz and from reactive to indigo dyed denim. The minimum shrinkage percentage was observed in 9 oz fabric, making it ideal for protective clothing.

Fabric Stiffness

The unfinished and finished state fabric did not show much of difference. The sample was less drapable when the fabric weight was high. Higher fabric weight and weight add-on resulted in this increase in stiffness but not more than 0.5 cm. Stiffness observed was less in 9 and 11 oz denim making it suitable for PC.

Crease Recovery

On application of finishes the crease recovery showed a slight difference of $0'$ to $5'$ degree angle. With increase in fabric weight there was decrease in the angle of recovery. Better results were observed in 9 oz denim.

Air Permeability

Table shows that the indigo dyed samples have higher permeability to air when the fabric weight was reduced. Readings for 9 oz has 9.3 cc/cm^2/s, 11 oz has 8.6 cc/cm^2/s, 13 oz has 7.2 cc/cm^2/s. 9 oz and 11 oz denim showed an increase in comfort factor to the wearer and therefore recommended for PC. The application of anti-microbial finish and flame-retardant finish has resulted in lowering of air permeability not more than 0.3 cc/cm^2/s.

Thermal Conductivity

Thermal conductivity was higher for 9 oz denim, that is, 0.17 m^2 (class 2) while 0.21 m^2 and 0.25 m^2 (class 3) for 11 and 13 oz indigo dyed denim due to lower thread count and weight of cotton. There was a marked decrease in 0.01 m^2 in all the samples of reactive dyed denim. Application of finish resulted in constant thermal conductivity. Table 4.16 B observe tog value for 9 oz denim was acceptable, ideal for warm conditions as in the spray paint and other industries.

Water Vapor Permeability

B120 sample showed better result 824.0 g/m^2/day. In general, there was increase in permeability from 9 to 13 oz samples due to increase of cotton content in the fabric. 13 oz indigo dyed denim resulted in better permeability. But ideal fabric for PC would be 11 and 9 oz, while 13 oz absorbs more water/moisture, and moisture takes time to evaporate, thus making the denim heavier.

Tensile Strength and Percent Elongation

Indigo dyed 13 oz denim due to higher fabric weight and twill structure had the most suitable tensile strength. 9 and 11 oz gave good average results. Increase in fabric weight resulted in increased compactness of the twill weave construction, hence resulting in better strength to various tension.

Water Absorbency

Water absorbency is the ability of denim to absorb water within its structure. 9 oz denim showed the maximum absorbency by 2 s and 11 oz showed absorbency by 3 s. This is due to capillary action and less surface area in 9 oz and 11 oz denim as compared to 13 oz denim. Hence, 9 and 11 oz denim were suitable for upper garments (shirt, jacket) in protective clothing as they easily absorb the water moisture and evaporates from body.

Tear Strength

The maximum tear strength was found in 13 oz indigo dyed denim, when value was 80.53 N in control sample and increase by 1 to 2 N. On the other hand, 9 and 11 oz gave average values. Hence 13 oz denim was ideal for lower garments in protective clothing and can be used in high tension, to resist the possibility of tear along the breach or damaged position.

Abrasion Test

The ability of fabric to resist surface wear caused by flat rubbing in contact with another material was observed as above average in all the three weights of indigo dyed denim.

Durability to Laundry

Results showed an increase in water absorption and permeability after laundry while the effect of anti-microbial properties did not wear out much in indigo dyed fabric. Hence, the fabric results in better linkage with finishes. Both the finishes after wash cycles gave similar results, that is the flame-retardancy was found to be good and stain release gave grade 4 out of 5 even after 10 wash cycles.

3.5 Selection of Appropriate Finish on Protective Clothing for Petrochemical Industry

The chemicals selected in the research process were ecofriendly, sustainable, and feasible in terms of cost and availability. Also, according to research on indigo dyes and current course of research results, it was found that the use of indigo dyed fabric would make it essential for long-term usage. Application of anti-microbial finish on indigo dyed denim ensures additional resistance to microbes.

3.6 Flame Retardant Property

The qualitative assessment of flame retardancy of the denim samples was carried out. The flame-retardant properties were compared with the control sample here, and it was found that the finished sample bare excellent results in all the weights of fabric. As far as fire catching was concerned as per ISO 15025, all the combinations exhibited an excellent rating with no flaming on vertical or horizontal edge, no after flame, nil in afterglow spread, no occurrence of debris observed, and hence, no debris that ignites the filter paper nor hole development in multilayer specimen.

3.7 Selection of Appropriate Finish on Protective Clothing for Welding Industry

From the preliminary experimental work for anti-microbial, flame-retardant finish in the last section emerged as the optimum option followed in the protective clothing for welding. During the course of work, it was found that finish applied for moisture management had to be withdrawn from combinations because of the decrease of water vapor permeability value in the results. The presence of moisture management finish with organo-phosphorous compound rendered not much absorption of moisture/water vapor. Thus, moisture management finish was not used for further study. In this view of development, the next finish from comparative assessment of three anti-microbial and one flame retardant finish was taken up.

Thus, it was evident that the combination anti-microbial [Sanigard 7500 (L.N. Chemical), Fabsheild AEM 5700 and Microban PB 80150–050 (Rossari Biotech Limited)] and flame retardant (organo-phosphorous compound, i.e., Glogard CP new); anti-microbial and stain release [Aquarepel XC-630 (L.N. Chemical)] offer a good option for a durable anti-microbial and flame retardant; anti-microbial and stain, oil release finish applicable on 100 % cotton denim fabrics. Addition of Fixer ALN to the finishes improves its durability to laundry. The chemicals used do not adversely affect the other properties. The garments were found to be effective against microbe penetration, flame or fire spark penetration, and stain or oil releasing, as shown by the laboratory results. Special design feature was incorporated to ensure adequate protection against hazards and better carrying capacity for tools and other things. The use of protective clothing is essential for the protection of workers in the process and OEM industries. Use of protective clothing developed is also recommended for workers in any other chemical industries.

4 Conclusion

The salient findings of the experimental work undertaken for the selection of appropriate finish and preparation of the finishes applied on denim were as follows:

- 100 % cotton denim and its 3/1 twill weave structure gave good durability which was observed to be long-lasting as compared to the blends with polyester and/or spandex that are currently and easily available in the market.
- Indigo dyed samples proved to give inherent anti-microbial properties that were long-lasting and better than that of reactive dyed samples.
- Denim washes for aesthetic appeal may reduce the amount of indigo dye in the fabric, which may or may not reduce the anti-microbial effect; hence, anti-microbial finish application was made primary in the research, resulting in 99.99 % resistance and more than 85 % resistance even after 5 wash cycles.
- Application of two finishes opposite in nature (moisture management and flame retardant) resulted in lowering of water vapor permeability. Hence, considering moisture management as one of secondary finish was not accepted.
- It was observed that the flame-retardant property and moisture absorption was better without moisture management finish, probably due to use of 100 % cotton denim fabric.
- Application of finishes on 9 and 11 oz resulted in most suitable values for protective clothing regarding comfort and textile handling properties.
- Other textile properties such as air permeability, thermal conductivity, tensile strength, tear strength, and abrasion resistant increased significantly in 13 oz, which made the denim heavier.
- Incorporation of Fixer ALN in flame retardant and stain release finish improved its durability to laundry up to 20 wash cycles, which is recommended as per the Indian standards.
- The evaluation of protection of finished fabrics against test methods was carried out using standards IS, ISO and AATCC. A marked improvement in every property was observed.

References

1. Bajaj P, Sengupta AK (1992) Protective clothing—Textile Progress, vol 22. Textile Institute
2. DeJonge JO, Easter EP, Leonas KK, King RM (1985) Dermal exposure related to pesticide use. Protect Apparel Res:403–411
3. Mistry KU (2012) Fundamentals of industrial safety and health. Siddharth Prakashan, Ahmedabad
4. A short history of denim. www.levistrauss.com/wp-content/uploads/2014/01/A-Short-History-of-Denim2.pdf.Accessed 22 June 2018
5. Vijayalakshmi D, Ramachandra T (2013) Application of natural oil on light weight denim garment and analysis of its multifunctional performances. Indian J Fibre Text Res 38:309–312
6. Bajaj P (2000) Heat and flam protection. In: Horrocks AR (2000) Handbook of technical textiles. Woodhead Publishing Ltd. Cambridge, pp 223–263
7. Henry MS (1980) Users' perceptions of attributes of functional apparel
8. McCann J (2005) End-user based design of innovative smart clothing. University of Wales Newport, UK, Wales
9. Suri M, Rastogi D, Khanna K (2002) Development of protective clothing for pesticide industry. Indian J Fibre Text Res 27:85–90
10. Ahmad I, Rehan M, Balkhyour M, Abbas M, Basahi J, Almeelbi T et al (2016) Review of environmental pollution and health risks at motor vehicle repair workshops challenges and perspectives for Saudi Arabia. Int J Agric Environ:1–23
11. Ganesan P, Sasikala L (2017) Functional finishes for apparels. https://www.fibre2fashion.com/industry-article/2371/functional-finishes-for-apparels. Accessed 22 Dec 2017

Development of Ecofriendly Multifunctional Textiles Using Peppermint Oil

Prachity Wankhade, Neha Mehra, and Vijay Gotmare

1 Introduction

Traditionally, textiles were considered as low technology domain as their primary functions are protection of modesty, providing microclimate and good look. With the intensification of global competition, textile manufacturing companies from developed countries are competing for a significant share of the global market by developing new technologies or new products. Companies are trying to differentiate their products with specific and special functions. The consumers are demanding textile products with higher performances, even in the "traditional" clothing and home textiles areas. Therefore, researchers have made many attempts to impart more functional characteristics to fabrics (in addition to the above-mentioned primary functions) so as to get what are known as multifunctional textiles [1]. Textile industry is a major contributor in polluting the environment. Besides generating harmful toxins during the manufacturing of fabrics, the industry also emits poisonous gases like methane into the air. Tons of textiles are produced for global population that continues to grow and compete for improvement in living standards. Consequently, tons of textile waste are generated every year which pollutes water, air and the sites where the waste is dumped in landfills. Chemical processing sector of the textile industry contributes majorly to the water pollution.

With growing environment consciousness, people are desirous of shifting toward ecofriendly products. Scientists and researchers are finding novel ways to find solutions to this problem by proposing the use of natural fiber sources, natural oils and natural herbal extracts for finishing of fabrics. Few oils are already being used for their antibacterial, antiodor, mosquito repellent, UV protection and for their aromatic properties. The use of harmful chemicals needs to be replaced by these natural products, thus leading to a sustainable way for health and environment benefits. Improvement

P. Wankhade · N. Mehra (✉) · V. Gotmare
Veermata Jijabai Technological Institute, Matunga 400019, India
e-mail: nehamehra@tx.vjti.ac.in

A. Majumdar et al. (eds.), *Functional Textiles and Clothing 2020*,
https://doi.org/10.1007/978-981-15-9376-5_11

of existing properties and the creation of new material properties are the most important reasons for the functionalization of textiles. Keeping this emerging trend in view, the present research work has been undertaken to explore various traditional materials, methods and the latest application techniques being used for multifunctional finishing.

A thin film fabrication technique, known for its simplicity in methodology as the LBL technique, is used for surface modification. Strong electrostatic attraction between the charged substrate and an oppositely charged molecule in the other solution leads to deposition of polyanions and polycations on the fabric surface. This phenomenon has long been known to be a factor in the adsorption of small organics and polyelectrolyte. But it is rarely observed in molecular layer formation. The accumulation of multilayers on solid substrate by layer by layer technique is a unique novel approach for the fabrication of ultrathin films on solid substrate [2]. This technique has been used earlier for the application of chrysanthemum oil nanoemulsion on nylon net fabric for mosquito repellent activity [3]. The application of titanium dioxide nanoparticles by layer by layer technique has also been reported on nylon fabric for its antibacterial property [4]. One of the recent trends in textile industry is "nanotechnology" which can provide high durability for fabrics as they have a large surface area to volume ratio and high surface energy, thus presenting better affinity for fabrics, leading to an increase in durability of the finish [5]. Keeping this emerging trend in view, nanotechnology and layer by layer application technique can be used to develop multifunctional products.

2 Materials

2.1 Textile Substrate

120 GSM plain woven scoured and bleached cotton fabric with EPI: 120, PPI: 80 and 40s count in the direct system, supplied from Piyush Trading, Mumbai, India has been used for the experimentation.

3 Chemicals

Peppermint oil has been procured from Shreeji Aroma Enterprise, Mumbai, India. Sorbitan monooleate (Span 80, chemically pure grade) as a lipophilic surfactant and polyethylene 20 sorbitan monooleate (Tween 80, of chemically pure grade) as a hydrophilic surfactant used as surfactant in oil in water nanoemulsion preparation were supplied by Croda Chemicals Pvt. Ltd. India. Cationic and anionic polyelectrolytes, polyethyleneimine and polyacrylic acid of Sigma Aldrich were procured from Rarco Research Lab, Mumbai, India.

4 Methods

4.1 Fabric Pretreatment

The cotton fabric has been treated with 2 g/l of non-ionic soap solution and distilled water at 60 °C for 30 min to remove any leftover dirt on the untreated fabric. Then the material is given hot wash, cold wash and air dried.

4.2 Preparation of Peppermint Oil in Water Emulsion

The preparation of O/W emulsion has been done based on the hydrophile and lipophile balance system. To make trial emulsions, matched pair of Span 80 and Tween 80 has been selected as they belong to the same chemical class "Oleates" [6].

The batch number from 3 to 7 in Table 1 has their calculated HLB in the range of 8–18, which are classified as oil in water emulsifiers. An example of determination of calculated HLB for a pair of emulsifiers is demonstrated as follows:

$$\begin{aligned} \text{SPAN80}: \quad & 85\,\% \times 4.3 = 3.66 \\ \text{TWEEN80}: \quad & 15\,\% \times 15.0 = 2.25 \\ & \text{HLBofblend} = 5.91 \end{aligned}$$

.

The emulsifiers described in Table 1 (batch 1 to 7) have been evaluated to provide stable O/W nanoemulsion of 6 % oil in water. The S_{mix} ratio of batch number 5, which gives a stable nanoemulsion without any flocculence, coalescence, sedimentation and foaming has been considered for the preparation of stable nanoemulsion. The nanoemulsion has been prepared using oil/S_{mix} ratio of 5.5:4.5. The measured quantity of emulsifiers and oil has been emptied into a glass beaker, which has been homogenized at 6000–8000 rpm, with the aid of a digital high-speed homogenizer.

Table 1 Emulsifier combinations for preparing O/W emulsion and their calculated HLB

Batch No	Span 80 (HLB 4.3)	Tween 80 (HLB 15)	Calculated HLB
1	100	00	4.3
2	85	15	5.91
3	67	33	7.83
4	45	55	10.19
5	28	72	12
6	06	94	14.4
7	00	100	15

The homogenization has been done for 30 min, and 5 min of break has been given after every 10 min of stirring to allow the particles to settle.

4.3 Nanoemulsion Application Through Layer by Layer Technique

Preparation of polyelectrolyte solution: Two polyelectrolyte solutions are prepared with the aid of a magnetic stirrer at room temperature. These polyelectrolytes are used in the LBL application technique, where oil nanoemulsion will be attached on fabric due to surface ionic bonding. The first solution contains 0.1 % polyethyleneimine (cationic polyelectrolyte), branched having molecular weight 25,000. The pH of polyelectrolyte solution is adjusted to 7 using 1 M HNO_3. The second solution contains 0.1 % polyacrylic acid (anionic polyelectrolyte) having molecular weight 450,000. The pH is adjusted to 5 using 1 M HNO_3 [4].

PAA contains carboxylic acid repeat unit that dissociates to form negatively charged anion in low pH aqueous solutions. With the increase in pH, the polymer chain becomes increasingly ionized, the building negative charge constrains further deprotonation, and the p*K*a, the dissociation constant value of PAA, alters. PAA adopts a relatively smooth swelling process in the pH range of 4–6. As the degree of ionization increased from pH 4 to 6, the equivalent deprotonation and subsequent anionic charge drive PAA to adopt an extended state with a relatively smooth transition, with only small changes to polymer physical properties that save additional anionic potential [7]. Thus, the pH of polyacrylic acid is adjusted to 5.

Application of Oil Nanoemulsion through LBL Technique: Fabric is immersed into a beaker containing a solution of cationic polyelectrolyte which results in activation of functional groups on the outer surface, that is, attachment of first layer. Cotton fabric has negative charge when immersed in water, thus when it is first dipped in the solution of cationic solution, positive charges from the cationic solution would easily bond with the negative charge on the cotton substrate. Thus, the fabric has been first dipped in the solution of positive polyelectrolyte solution for 2 min. A washing step with distilled water is followed to remove the loosely adhered polyelectrolyte. The fabric is then dipped into a beaker containing solution of negative polyelectrolyte and oil nanoemulsion for 2 min. This is again followed with a washing step to remove the loosely deposited electrolytes. The cycle is repeated till the required number of layers are obtained. The number of layers is the number of times the cycle of dipping the fabric into alternate solutions of polyelectrolyte is repeated. When the fabric is dipped into alternate solutions of polyelectrolytes, multilayers are spontaneously deposited on the fabric surface through strong electrostatic formation. The finished sample is air dried at room temperature and stored in an airtight bag [3].

Trials have been conducted using 50 and 100 g/l of nanoemulsion concentration in the negative polyelectrolyte solution. The nanoemulsion is added in the negative polyelectrolyte solution as the oil would easily get attached from the anionic solution

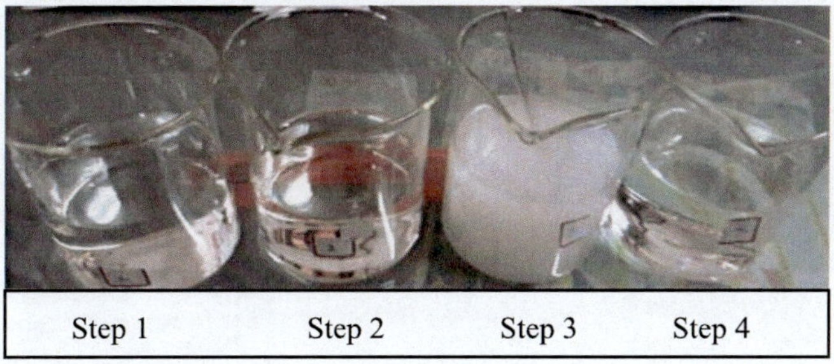

Fig. 1 LBL technique lab setup

to the cationic polyelectrolytes on the substrate. The demonstration of the LBL technique is shown in Fig. 1.

4.4 Testing of Fabric for Its Multifunctional Properties

It is important to analyze the size and stability of the nanoemulsion and various properties incorporated on the fabric by the treatment. The characterization of nanoemulsion has been done on the Malvern Mastersizer 2000 particle size analyzer which uses laser diffraction techniques to measure the size of the particles. A mathematical model (Mie or Fraunhoffer Theory) is applied to generate a particle size distribution. The final result is reported on an equivalent spherical diameter volume basis.

The antimicrobial activity efficiency of the finished cotton fabric sample against *Staphylococcus aureus* strain no. ATCC 6538 (gram-positive bacteria) and *Klebsiella pneumoniae* strain no. ATCC 4352 (gram-negative bacteria) has been quantitatively done by AATCC 100–2012.

UV-protection factor (UPF) has been determined using UV transmittance analyzer UV 2600 according to the Australian/New Zealand standard (AS/NZS 4399).

Mosquito repellence has been evaluated using modified WHO/CTD/WHO PES/IC/96.1 test where an excito repellency chamber is used. Laboratory-reared mosquitoes are released in an excito repellency chamber. The chamber consists of two cubicle chambers attached together with a hole in the central wall. The right side of the chamber walls is covered with finished sample and the left side of the chamber walls is covered with unfinished fabrics. Mosquitoes are released to the right side of the chamber to observe changes in behavior in the form of moving away from treated fabric to untreated fabric through the hole in the central wall.

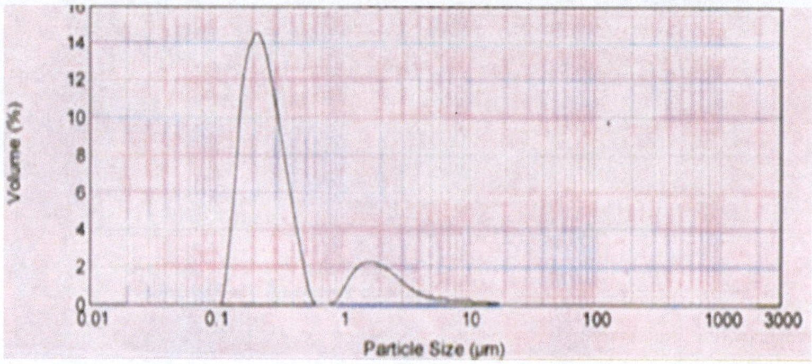

Fig. 2 Particle size distribution of oil nanoemulsion

5 Results and Discussions

5.1 *Particle Size Analysis of Peppermint O/W Nanoemulsion*

The characterization of nanoemulsion has been done on the Malvern Mastersizer 2000 particle size analyzer which uses laser diffraction techniques to measure the size of the particles. The particle size of emulsion has been measured after storing the sample at 27 °C in an incubator for 30 days to check the stability. The mean diameter of the total particles of nanoemulsion has been found to be 0.265 μm and the volume weighted mean has been found to be 0.647 μm, which indicates that the oil droplets forming the emulsion are in the nanometer range and a stable emulsion is formed. The graph in Fig. 2 shows that 5 % of the total particles of emulsion has a mean diameter of 0.162 μm, 10 % of the total particles of emulsion has a mean diameter of 0.612 μm and 50 % of the total particles of emulsion has a mean diameter of 0.254 μm. Figure 2 demonstrates a narrow size distribution which depicts an ideal mean size distribution according to the Mastersizer 2000 (Figs. 3 and 4).

5.2 *Physical Testing of Untreated Cotton Fabric*

The cotton fabric has been assessed for its physical and chemical properties to ensure whether the fabric is suitable for its end application of the finishing formulation. The EPI and PPI of the cotton has been found to be 120 and 80, the warp and weft count has been found to be 40 s, the calculated GSM is 126.43, the breaking strength warp way and weft way has been found to be 105 and 30 kgf and the abrasion resistance has been found to be 3.8 % in terms of loss in thickness and 2.6 % in terms of loss in weight. The fabric has been found to burn readily in flame, emitting burned paper like smell in the burning test and has dissolved in 70 % H_2SO_4 + boil, thus confirming cotton fiber. From the experimental analysis it is clear that cotton

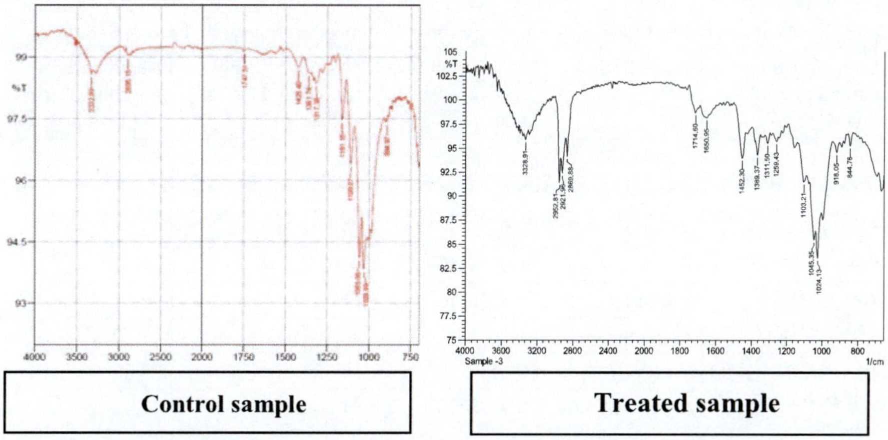

Fig. 3 FTIR analysis of untreated fabric and fabric treated with 100 g/l per 20 l of peppermint oil nanoemulsion through LBL technique

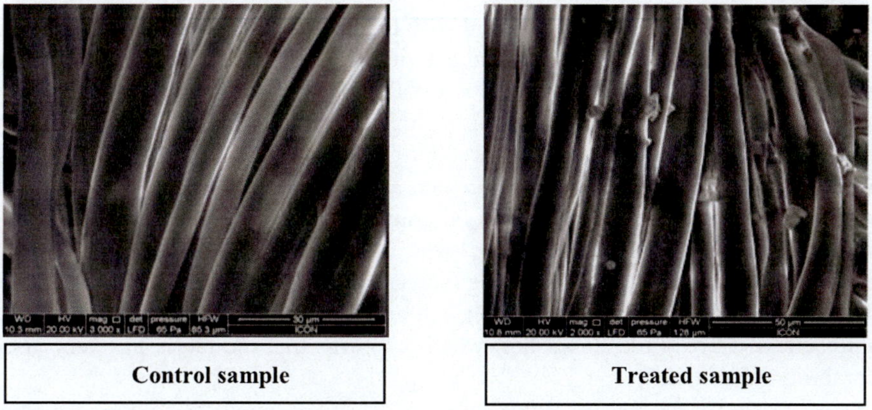

Fig. 4 SEM images of control sample and sample treated with 100 g/l per 20 l of peppermint oil nanoemulsion through LBL technique

sample shows significant performance where durability is concerned as the measure of abrasion and breaking strength is good, thus ensuring the fabric is suitable for the end finishing.

Table 2 Antibacterial test results of samples treated with peppermint oil by LBL technique

Emulsion conc. and number of layers	Test culture	No. of colonies recovered at "0" hour [B]	No. of colonies recovered at "24" hours [A]	Reduction of microorganisms [R] in %
Control	*S. aureus*	2.04×10^5	2.12×10^7	0
	K. pneumoniae	2.08×10^5	2.16×10^7	0
50 g/l/20 l	*S. aureus*	1.92×10^5	1.2×10^4	93.7
	K. pneumoniae	1.99×10^5	1.7×10^4	91.45
100 g/l/20 l	*S. aureus*	1.52×10^5	3×10^3	98.02
	K. pneumoniae	1.76×10^5	4×10^3	97.72
100 g/l/20 l and 20 cycles washed sample	*S. aureus*	0.65×10^5	1.1×10^4	83.07
	K. pneumoniae	0.70×10^5	1.3×10^4	81.42

5.3 *Antibacterial Test*

Antibacterial testing of fabric treated with oil nanoemulsion through LBL technique: Table 2 shows the antibacterial activity of 120 GSM cotton woven scoured and bleached sample which is treated with 50 and 100 g/l of oil nanoemulsion and number of layers employed in the LBL technique are 20 (100 g/l/20 l). The sample treated with 100 g/l/20 l showed highest antibacterial efficiency of 98.02 % against *S. aureus* and 97.72 % against *K. pneumoniae*. The antibacterial efficiency of fabric sample increased with the increase in oil emulsion concentration. The strong ionic bonds formed on the cotton surface due to the use of polyelectrolytes in the LBL technique have deposited more oil with the increase in nanoemulsion conc. and number of layers. The fabric treated with 100 g/l of emulsion has been found to possess its antibacterial activity till 20 washes, which confirms that the fabric has very good durability of the finish.

5.4 *Mosquito Repellent Testing*

Mosquito repellent testing of fabric treated with oil nanoemulsion through LBL technique: Table 4 shows the mosquito repellency rate of fabrics treated with 50 and 100 concentration of peppermint oil nanoemulsions with 10 and 20 number of layers used in the LBL technique. The treated fabrics were subjected to excito repellency test. The fabrics treated with 100 g/l of peppermint oil nanoemulsion and 20 number of layers gave 100 % mosquito repellency rate. The fabric has been found to possess its mosquito repellency till 20 washes which confirms that the fabric has good durability of the finish.

Table 3 Mosquito repellency results of samples treated with peppermint oil by LBL technique

Emulsion concentration and number of layers	No. of mosquitoes released in treated chamber	No. of mosquitoes on treated fabric	No. of mosquitoes on untreated fabric	No. of mosquitoes showing mobility	Percentage repellence
Control	10	10	0	0	0
50 g/l per 10 l	10	7	3	0	30
50 g/l per 20 l	10	6	4	0	40
100 g/l per 10 l	10	1	9	0	90
100 g/l per 20 l	10	0	10	0	100
100 g/l per 20 l and 20 cycles of washed sample	10	3	7	0	70

Table 4 UPF rating

S. no	Sample identification	UPF (290–400 nm)
1	Control	4.77
2	Sample treated with 100 g/l/20 L	7.28
3	100gpl/20 L and 20 cycles washed sample	6.55

5.5 UV Protection Factor Testing

Peppermint oil has been found to have UV protection factor ranging from 6 to 8. The fabric sample treated with oil emulsion by LBL technique showed UPF factor to be 7.28. The sample has been found to possess UPF 6.55 after 20 wash cycles. This indicates the fabric sample possess moderate UV protection ability.

5.6 FTIR Analysis of Fabric Treated with Oil Nanoemulsion Through LBL Technique

The FTIR has been done on FTIR 8400S spectrophotometer to detect the presence of functional groups. Peppermint oil contains isomers of cyclic ketones. A peak at 1714 cm^{-1} has been observed due to stretching of six-membered cyclic ketone in the FTIR graph. The FTIR analysis of treated fabric shows a peak at 1714 cm^{-1}, thus confirming the presence of peppermint oil.

5.7 SEM Analysis of Samples

The SEM images of two samples are depicted in Fig. 5. One is control sample which is scoured with non-ionic soap solution and the other is sample treated with peppermint oil nanoemulsion through layer by layer technique. The difference in the SEM images of control and treated sample clearly represents the nanoemulsion traces on the cellulosic surface of cotton fabric.

6 Conclusion

Stable oil in water nanoemulsion of peppermint oil was successfully prepared with the help of a hydrophilic and lipophilic surfactants. The thin film fabrication process of LBL brings about activation of functional groups on the cotton surface, thus attaching multilayers on the fabric through ionic bonding. The increase in oil nanoemulsion concentration and number of layers in the LBL technique has been found to boost the multifunctional properties of cotton fabric. The electrostatic bond formation due to the employment of polyelectrolytes in the LBL technique has led to an increase in oil deposition on the cotton fabric surface, thus improving the durability of the finish.

References

1. Kathirvelu S, Souza L, Dhurai B (2008) Comparative study of multifunctional finishing of cotton and P/C blended fabrics treated with titanium dioxide/zinc oxide nanoparticles. Indian J Sci Technol 1(7):1–12
2. Dey D, Islam MN, Hussain SA, Bhattacharjee D (2008) Layer by Layer (LbL) Technique for fabrication of electrostatic self-assembled ultrathin films 4(1):39–44
3. Bhatt L, Kale RD (2015) Development of mosquito repellent textiles using chrysanthemum oil nanoemulsion. Int J Text Fashion Technol 5(3):15–22
4. Kale RD (2013) Synthesis of Titanium dioxide nanoparticles and application of nylon fabric using Layer by Layer technique for antimicrobial property. Adv Appl Sci Res 3:3073–3080
5. Ratner M, Ratner D (2002) Nanotechnology: a gentle introduction to the next big idea. Prentice Hall PTR, New Jersey
6. ICI Americas Inc (1980) The HLB system, a time saving guide to emulsifier selection. Chemmunique, Wilmington, Delaware
7. Thomas S (2019) pH dependence of acrylate-derivative polyelectrolyte properties, acrylate polymers for advanced applications

Sustainable Production and Supply Chain

Reuse of Cigarette Filters for Energy Applications

Prakash Giri, Ashish Kakoria, Sahil Verma, and Sumit Sinha-Ray

1 Introduction

Cigarette smoke consists of elements and metals such as As, Cd, Cr, Fe, Hg, and Pb [1, 2], and more than thousands of components are reported to be present in cigarette smoke [3–5]. Some of the components, including Pb, Cd, and Cr, are considered to be highly carcinogenic and are reported to be present beyond the acceptable limit [6]. Presence of carbon monoxide (CO), carbon dioxide (CO_2), hydrogen cyanide (HCN), formaldehyde, acrolein, acetaldehyde, ammonia, nitrogen dioxide (NO_2), sulfur dioxide (SO_2), and many aromatics amines like anilines, toluidines, ethylanilines, dimethylanilines, naphthylamines, and aminobiphenyls has been well reported [7, 8]. Tobacco smoke components are influenced substantially by the nature and concentration of compounds that constitutes its basic portions [8, 9]. Asthma and cardiovascular diseases are very commonly reported among smokers [4, 10]. Chemical constituents of complex smoke aerosol have been studied to cause adverse biological activities including increased risk of pulmonary carcinoma [11, 12]. Cigarette smoke condensate (CSC) from pyrolysis of tobacco proteins has been reported to enhance mutagenicity, influenced by aldehydes and ketones present in the vapor phase of cigarette smoke [11, 13, 14]. Patents and researches have reported the fibers of polypropylene, polyvinyl alcohol, and so on, and absorbents like carbon and silica to be effective materials for mainstream cigarette smoke filtration [15–17]. Cellulose acetate fibers mixed with other materials have been extensively used in commercial cigarettes.

P. Giri · A. Kakoria · S. Verma · S. Sinha-Ray (✉)
School of Engineering, Indian Institute of Technology Mandi, Mandi 175075, HP, India
e-mail: sumitsinha@iitmandi.ac.in

A. Majumdar et al. (eds.), *Functional Textiles and Clothing 2020*,
https://doi.org/10.1007/978-981-15-9376-5_12

2 Problem Statement

Toxicity of cigarette filters increases along with smoking, as significant amount of tar and chemical components gets absorbed [18]. These types of filters usually take up to 18 months to degrade normally, and are significant source of pollution for our surroundings and marine life [19]. Therefore, people eventually end up consuming a part of the toxic chemicals via biomagnification which otherwise are avoided in the mainstream cigarette smoke by filtration. Thus, it is crucial to restrict these filters to be exposed in the surrounding. Also, these cigarette filters are hardly reused and hence an alternative usage of these filter membrane, which contain carbon-rich elements, in terms of energy applications may open a plethora of options for further study. Here, in this work we have demonstrated the application of electrospun PAN nanofibers as efficient cigarette filter which can be further reused for energy storage application, thus completing a waste to energy nexus.

3 Methodology

3.1 Materials

Polyacrylonitrile (PAN) (mol. wt. = 150 kDa) was purchased from Sigma-Aldrich. The solvent of the polymer is N, N-dimethylformamide (DMF) which was purchased from Alfa-Aesar.

3.2 Electrospinning and Membrane Preparation

PAN was dissolved in DMF to prepare a 9 wt.% solution. The mixture was stirred at 40 °C and 300 rpm for 12 h. The solution was electrospun with a conventional laboratory assembled setup, the schematics of which is shown in Fig. 1. Solution was pumped through an 18-gauge needle at a flow rate of 1.2 mL/h using syringe pump purchased from New Era Inc. High voltage of 8 kV was supplied while maintaining the needle to collector distance of 12 cm. The collector was a rotating drum which was rotated at 100 rpm. The membrane was collected after 8 h of electrospinning. The physical image of the membrane is also shown in Fig. 1 (inset).

3.3 Characterization

Scanning electron microscopy (SEM) was used for the architectural study of filter membrane—before and after filtration of the filter membranes.

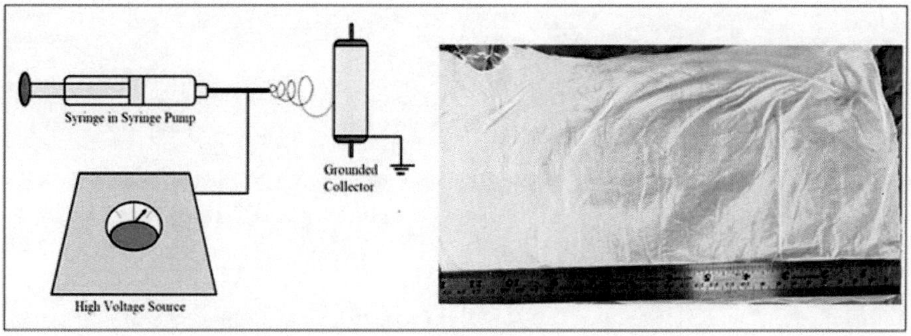

Fig. 1 Schematics of conventional electrospinning setup (left) and PAN nanofiber membrane prepared by electrospinning process (right)

3.4 Cigarette Smoke Filtration

Filter membranes produced by electrospinning process were rolled to form cylindrical filters of diameter 7.6 mm and length 15 mm to mimic a cigarette filter purchased from market for geometrical comparison. Degree of compactness in rolling process of the filter was fixed after trial-and-error basis to achieve similar molar flux as commercial cigarette filter in near similar pressure differential. Experimental scheme involved integrating manometer and flow meter to measure the pressure drop and flow rate through the filter membranes. Five measurements were taken for commercial cigarette filters and by replacing commercial cigarette filters with PAN filters for statistical parity and the average data is reported later in the *Results* section. Smoking was carried out with the help of squeezing and expanding action of a small rubber container to mimic human smoking.

3.5 Electrochemical Tests

Electrospun PAN nanofibers, both without smoking and after smoking, were carbonized for further electrochemical studies. The carbonization process was carried out at 900 °C in N_2 atmosphere after being stabilized in air at 300 °C. Soaking time of 1 h was maintained for both the processes at their respective atmosphere. Later the carbonized pan fibers were tested for their capacitance using cyclic voltammetry (CV) using Autolab Electrochemical Worsktation (Metrohm). For this, a conductive ink was prepared by adding 4 mg of carbonized PAN (separately for before and after smoking) and 1 mg of Vulcan carbon in a mixture of 25 μL Nafion and 500 μL of isopropyl alcohol. The mixture was sonicated for 30 min to get a homogenous solution.

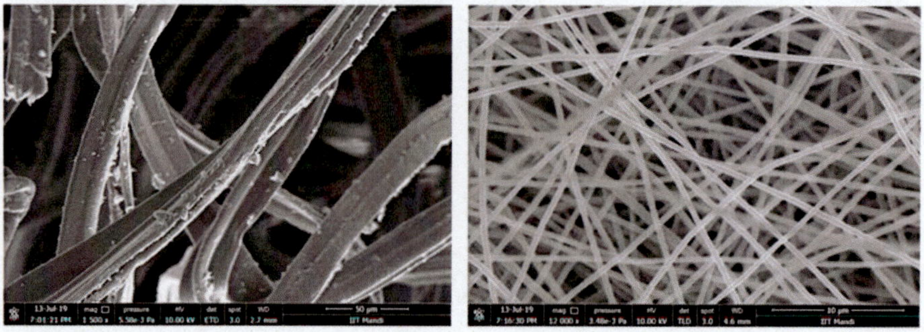

Fig. 2 SEM image of commercial cigarette filter (left) and PAN nanofibers (right). Scales are embedded in the images

4 Results and Discussion

4.1 Membrane Morphology

Cigarette filters of a specific brand were taken for experiments and were tested from the same batch. The individual commercial filter was 15 mm in length and 7.6 mm in diameter. The average weight of cigarette filter before filtration was 0.0792 g. Cigarette filters consisted of cellulose-based microfibers (25–30 μm) (c.f. Fig. 2). Initially, electrospun nanofiber flat membrane of PAN had a surface density of 5 GSM and thickness of 0.2 mm. PAN filters were prepared by rolling electrospun PAN nanofiber membrane. However, the PAN filters after rolling have similar dimension weighed much lesser (0.0121 g) than commercial filter. These fibers with smaller diameters are more suitable for filtration purposes as they have high porosity and high surface area to volume ratio along with increased tortuosity which allows ample opportunity for enhanced interception and adsorption [20].

4.2 Filtration Potential

Commercial cigarette filters and PAN nanofiber filters were subjected to filtration of smoke from different cigarettes of the same production batch. The average pressure drop in commercial cigarette filter was found to be 3.87 kPa with a molar flux of 2.3×10^{-3} mol/m^{2}s. The pressure drop and molar flux in PAN nanofiber filter were measured to be 4.16 kPa and 2.6×10^{-3} mol/m^{2}s, respectively, under the same testing condition (Fig. 3).

Commercial cigarette filters gained around 25 % of their initial weight whereas PAN nanofiber filters gained around 115 % of their initial weight after filtration. Energy dispersive spectroscopy (EDX) results of the filter membranes after smoke filtration showed the presence of various elements including heavy metals like Pb,

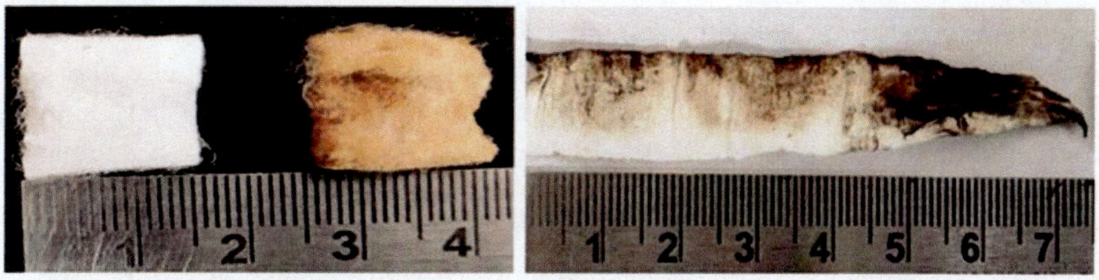

Fig. 3 Commercial cigarette filters before and after filtration (left) and PAN nanofiber membrane after filtration

Cd, and Cr. The filters were washed in 20 mL of DI water for 12 h at 30 °C and 300 rpm in a stirrer to leach the absorbed elements from the filters. The AAS results of the solution confirmed the presence of 11.55 ppm concentration of Pb in case of commercial cigarette filter leachate, whereas the concentration of 23.8 ppm was found in case of PAN nanofiber filter leachate.

4.3 Reusability in Energy Applications

The pure polymeric nanofiber membrane was initially non-conductive. However, post carbonization they became conductive, which make them suitable for current collection, as well as for electrocatalyst studies, since many transition elements were already doped in the PAN fibers, which should be retained in the CNFs.

A comparative analysis was made between two different CNFs obtained from PAN fibers, one pure PAN fibers converted to CNFs and the other PAN fiber filter converted to CNFs after smoking. In Fig. 4 it is visible that after carbonization, nanofibers from both parent PAN filter and post smoking PAN filters retained their fibrous and cylindrical form. The only difference in the later was their tendency to fuse with each other, which could have happened during smoking, where nanofibers

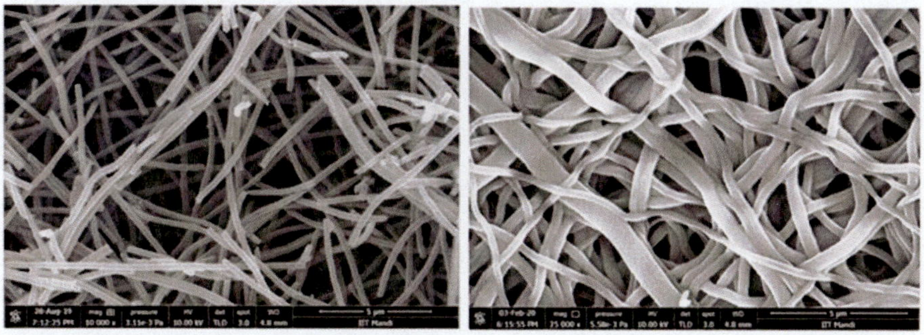

Fig. 4 Carbon nanofibers: without smoke filtration (left) and after smoke filtration (right)

could have gotten sintered with each other. It is known that PAN has glass transition temperature (T_g) ~ 80 °C [21], and given that temperature inside cigarette filters can easily go up to 100 °C [22], it is only possible that some nanofibers have sintered after attainment of T_g. However, visibly their nanofibrous architecture has remained the same. It was found that electrochemical activity of the membrane carbonized after filtration was increased, as revealed from CV (Fig. 5a). EDX confirmed the retention of various transition metals including previously stated Pb, Cr, Fe, and Cd even after carbonization (Fig. 5b). The presence of these elements is crucial as they act as dopants, thus increasing electrochemical activities [23].

The samples were oxygen reduction reaction (ORR), hydrogen evolution reaction (HER), and oxygen evolution reaction (OER) active. The pseudo-capacitance was increased by 202 % when membranes after cigarette smoke filtration were used as the test specimens. The specific capacitance was calculated based on the empirical relation [24]:

$$c = \frac{\int ivdv}{2\mu m \Delta V}$$

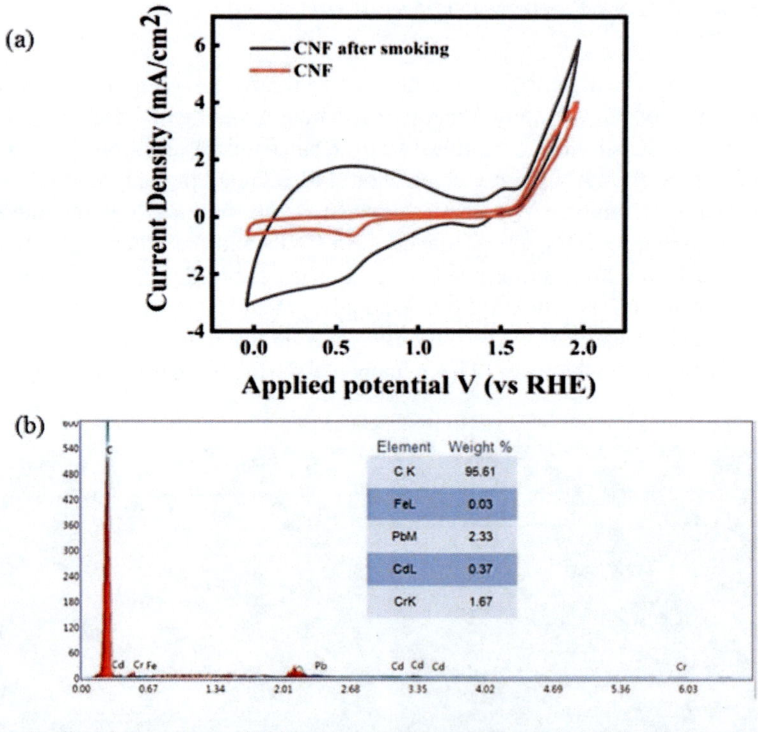

Fig. 5 **a** Cyclic voltammogram of CNFs prepared after carbonization of PAN membranes. Legends are embedded in the image; **b** EDX spectra of CNF obtained from post smoking filters.

where i and v were the current and potential in CV test, μ was the scan rate, m was the mass of active material, and ΔV was the potential window of discharge. The specific capacitances on the samples carbonized before and after filtration were found to be 138 and 417 F/g, respectively.

5 Conclusion

The major problem associated with management of used cigarette buds can be eliminated when these buds can be utilized for value-added products. An example with PAN nanofiber filters has shown that filters made out of these membranes can be carbonized after filtration and can be made electrochemically active to use them as current collector, catalyst or catalyst support material, and electrode material for fuel cells and battery manufacture. Besides, the cigarette smoke filtration performance is enhanced in nanofiber-based filters than commercial cigarette filters by about 4.3 times described by tar-based component retention capacity. Moreover, the absorption of heavy metals like lead and arsenic was also found to be increased by more than two times. The enhancement is due to nanofibers' high surface area to volume ratio than commercial cigarette filter's microfibers, which increases tortuosity of the architecture, allowing more fiber–smoke interaction, and thus making interception-based separation more efficient. Therefore, this study suggests that the cigarette filter membranes made out of nanofibers which can be converted to porous and carbon-rich fibers should further be researched. Here the reuse of efficient cigarette filter made out of PAN fibers in terms of electrochemical energy storage is one of many self-sustaining solutions of such filters. This study also provides an opportunity to use filters made up of PAN nanofiber membranes for automobiles and traditional HVAC requirements with end life usages for battery applications.

Acknowledgements The authors acknowledge Advanced Material Research Centre (AMRC), IIT Mandi for their advanced instrumentation facility. P.G., A.K. and S.V. acknowledge scholarship from Ministry of Human Resource Development (MHRD), India. S. S-R. Acknowledges financial support from DST-SERB (ECR/2017/001511)

References

1. Chiba M, Masironi R (1992) Toxic and trace elements in tobacco and tobacco smoke. Bull World Health Organ 70(2):269
2. Stohs SJ, Bagchi D, Bagchi M (1997) Toxicity of trace elements in tobacco smoke. Inhal Toxicol 9(9):867–890. https://doi.org/10.1080/089583797197926
3. Rodgman A, Smith CJ, Perfetti TA (2000) The composition of cigarette smoke: a retrospective, with emphasis on polycyclic components. Hum Exp Toxicol 19(10):573–595
4. World No Tobacco Day (2018) Tobacco breaks hearts – choose health, not tobacco. Geneva: World Health Organization; 2018 (WHO/NMH/PND/18.4). Licence: CC BY-NC-SA 3.0 IGO

5. Wright GF (1956) Studies with tobacco smoke condensate. In: Proceedings of the 3rd national cancer conference, pp 479–484
6. Hossain MT, Hassi U, Huq SI (2018) Assessment of concentration and toxicological (Cancer) risk of lead, cadmium and chromium in tobacco products commonly available in Bangladesh. Toxicol Rep 1(5):897–902
7. Kensler CJ, Battista SP (1963) Components of cigarette smoke with ciliary-depressant activity: Their selective removal by filters containing activated charcoal granules. New Engl J Med 269(22):1161–1166
8. Patrianakos C, Hoffmann D (1979) Chemical studies on tobacco smoke LXIV. On the analysis of aromatic amines in cigarette smoke. J Anal Toxicol 3(4):150–154
9. Schmeltz I, Hoffmann D (1977) Nitrogen-containing compounds in tobacco and tobacco smoke. Chem Rev 77(3):295–311
10. Karch A, GBD (2016) Risk factors collaborators. Global, regional, and national comparative risk assessment of 84 behavioural, environmental and occupational, and metabolic risks or clusters of risks, 1990–2016: a systematic analysis for the Global Burden of Disease Study 2016
11. Rodgman A, Smith CJ, Perfetti TA (2000) The composition of cigarette smoke: a retrospective, with emphasis on polycyclic components. Hum Exp Toxicol 19(10):573–595
12. Doll R, Bradford Hill A (1952) Study of the aetiology of carcinoma of the lung. Br Med J 2(4797):1271
13. Clapp WL, Fagg BS, Smith CJ (1999) Reduction in Ames Salmonella mutagenicity of mainstream cigarette smoke condensate by tobacco protein removal. Mutat Res/Gen Toxicol Environl Mutagen 446(2):167–174
14. Smith CJ et al (2000) "IARC group 2B Carcinogens" reported in cigarette mainstream smoke. Food Chem Toxicol 38(9):825–848
15. Buntin RR, Harding JW, Keller JP, Wollie L, Murdock AO (1971) Patent No. 3595245. United States
16. Xue LL, Bis.Koller K, Gao Q (2003) Patent No. US 6584979 B2. United States
17. AIP Conference Proceedings 1755, 150006 (2016) https://doi.org/https://doi.org/10.1063/1.4958579. Accessed 21 July 2016
18. Slaughter E, Gersberg RM, Watanabe K, Rudolph J, Stransky C, Novotny TE (2011) Toxicity of cigarette butts, and their chemical components, to marine and freshwater fish. Tobacco Control 20(Suppl 1):i25–i29
19. Mohajerani A, Kadir AA, Larobina L (2016) A practical proposal for solving the world's cigarette butt problem: recycling in fired clay bricks. Waste Manage 1(52):228–244
20. Brown RC, Wake D (1991) Air filtration by interception—theory and experiment. J Aerosol Sci 22(2):181–186
21. Jianhui W, Wen D, Bin P, Xiaobing Z, Fuwei X, Huimin L, Kejun Z (2014) Filtration and retention characteristics of smoke components in filters. Beiträge zur Tabakforschung International/Contrib Tobacco Res 1;26(3):121–131
22. Cox RT (2014) Inventor; Borealis Technical Ltd, assignee. Cooling filter for cigarettes and smoking articles. United States patent application US 13/853,034
23. Bosch-Jimenez P, Martinez-Crespiera S, Amantia D, Della Pirriera M, Forns I, Shechter R, Borràs E (2017) Non-precious metal doped carbon nanofiber air-cathode for Microbial Fuel Cells application: oxygen reduction reaction characterization and long-term validation. Electrochim Acta 20(228):380–388
24. Devi B, Venkateswarulu M, Kushwaha HS, Halder A, Koner RR (2018) A polycarboxyl-decorated FeIII-based xerogel-derived multifunctional composite (Fe_3O_4/Fe/C) as an efficient electrode material towards oxygen reduction reaction and supercapacitor application. Chemistry–A Eur J 2;24(25):6586–6594

Developing Organic Fabric from Aquatic Cellulosic Waste

Madhu Sharan and Sumi Haldar

1 Introduction

The textile industry across the globe is constantly striving for an innovated supply chain to gain sustainable development in the entire sector. It is high time to reduce the size of the environmental footprints made by this sector. This includes the development of sustainable and eco-friendly raw materials, less energy, intensive and minimum polluting process sequences and technologies and so on. Cellulosic materials are the cheapest and most abundant renewable resource in nature. Natural cellulosic fibers are obtained from the stem, leaf or seeds of the plants and all have numerous applications. There is an urgent need to manufacture a textile product from the renewable resources such as plants/trees (bast and leaf fibers) produced by ecologically sound manufacturing processes, transported and used with minimum environmental burdens, and finally ending its life smoothly without any additional environmental burdens (recyclable and biodegradable at the end of the life) [1].

India has a rich variety of wetland habitats due to varied topography and eco-climatic regimes. Types of wetland plants are also remarkably diverse owing to the large-scale variation in topography, huge climatic gradient across the country, and due to the local variation in rainfall. All wetland plants have been grouped under following broad categories that is free-floating hydrophytes (*Eichhornia crassipes, Lemina perpusilla, Pistia stratiotes*, etc.), suspended hydrophytes (*Hydrilla* sp., *Utricularia* sp., etc.), submerged anchored hydrophytes (*Aponogeton appendiculatus, Vallisneria* sp., *Potamogitum* sp. etc.), anchored hydrophytes with floating leaves (*Nelumbo nucifera, Aponogeteus natans,* etc.), anchored hydrophytes with

M. Sharan · S. Haldar (✉)
Department of Clothing and Textiles, Faculty of Family and Community Sciences, The Maharaja Sayajirao University of Baroda, Vadodara, Gujarat, India
e-mail: sumihaldar924@yahoo.com

M. Sharan
e-mail: madhusharan@yahoo.co.in

A. Majumdar et al. (eds.), *Functional Textiles and Clothing 2020*,
https://doi.org/10.1007/978-981-15-9376-5_13

floating shoots (*Ipomoea aquatica, Trapa maximowiczii*) and emergent anchored hydrophytes (*Bacopimonnieri, Eleocharisspiralis, Hydroceratriflora, Linnocharis-flava, Limnophila aromatic, L. heterophylla, Monochoriavaginalis, Typha* spp. etc.) [2].

Lotus is the ancient aquatic plant that is widely distributed across the globe in China, India, Bhutan, Malaysia, New Guinea, Japan, Pakistan, Philippines, Russia, Sri Lanka, America, Australia and Thailand [3]. Fibers of Nelumbo nucifera Gaertn, an aquatic perennial plant more commonly known as lotus with its pink and white flowers, are sacred in parts of Asia. It is also reported that the stems of pink color lotus submerged in the water are perfect to produce strong and durable fiber [4]. Fiber-producing species Nelumbo nucifera Gaertn occurs widely from Kashmir to Kanyakumari showing huge phenotypic diversity with different shapes, sizes and shades [5]. Different parts of Nelumbo nucifera Gaertn seeds, rhizomes, flowers, stamens and roots have numerous applications. The flowers have a most economic importance across the country used in festive occasions and offerings to god and goddess. Due to the wide usage of flowers, the petioles are cut with the little length of petiole leaving the rest of the entire petiole in the pond as a waste. This petiole contains precious fibers that are hidden in the water. People often confuses between the petiole and rhizome. In normal colloquial language people speak petiole as a stem. But scientifically both the parts are different. The petiole is the stalk that supports a leaf in a plant and attaches it to the stem. Many people often call it a stem, which is incorrect. A stem is the part of the plant that serves as the main source of support and produces nodes and roots, and that's not what we observe in petioles [6]. Rhizome (also known as rootstocks) is a type of plant whose stem is situated either at the soil surface or underground that contains nodes from which roots and shoots originate (shown below). Rhizomes are unique in that they grow perpendicular, permitting new shoots to grow up out of the ground. When separated, each piece of a rhizome is capable of producing a new plant [7].

The lotus flower is adorned for its characteristic of rising above the muddy water, indicating how one can rise above defilements of life. Apart from motivation for life, the plant also provides fibers which are used for making a rare kind of cloth matching with the flawless virtues of the silk. Fibers extracted from the lotus flowers of the Myanmar lakes are spun by hand and woven within 24 h making a fabric similar to silk. Extracting fibers from lotus stems has been in practice since 1910. Later during the 1990s designers of Japan setup workshops to create a foreign market for their fabric. But due to low demand in Japan, lotus fiber fabric remained a rare and handmade textile [1].

India is the country where lotus farming is practiced nearly about in all the states. Lotus cultivators used to sell the flowers, leaves and rhizomes but they are not aware about the petioles which contain precious fibers that are left in the pond after picking the flowers. As per the review, the manual extraction pattern followed in the Myanmar is: first, the fibers are spun by palm twisting, and again it is taken on the hand charkha which is time-consuming. In the present study researcher has tried to extract the fiber from the petiole and directly winded in the pern without palm twisting. The fibers were handspun in the box (*peti*) *charkha* and fabrics were prepared. The aim of the research was to experiment with the lotus petiole waste for its extracted fibers and test its properties, spinnability of the extracted fibers on box *(peti) charkha,* and construct union fabrics on handloom.

2 Experimental

2.1 Procurement of Lotus Petioles

For the procurement of lotus petiole, the local lotus flower vendors of Khanderao market, Vadodara were initially contacted (Figs. 1 and 2). First few samples were collected from them. Further for larger quantity the vendors suggested and helped in contacting the cultivators in and around Vadodara district who could provide petioles in bulk. The cultivators were contacted and convinced to provide petiole as and when in required quantity. Species used in this research was Nelumbo nucifera Gaertn. (pink form) belonging to the family Nelumbonaceae. The length of the petiole varies

Fig. 1 Procurement of lotus petioles

Fig. 2 Lotus petioles

from 50 to 115 cm and color varies from light to dark green and spikes were seen on the surface of petiole.

2.2 Fiber Extraction

Petioles were washed and wiped. It was not possible to extract fibers from single/two petioles, so a bunch of three petioles was taken and slit into 5–7 sections as per its length with the help of a sharp knife. After one slit a bunch of three petioles was slowly stretched and fibers were laid on a wooden slab. End-to-end points were joined by palm twisting and the yarns were wounded on the pern.

2.3 Fiber Testing

Single unit: Fibers taken from one cell (hole) are considered as a single unit, as shown in Fig. 3. Inside the lotus petiole there are two types of xylem cells: tracheids and vessels. Tracheids are elongated narrow tube-like cells with hard thick and lignified walls with large cell cavity. Vessels are cylindrical tubular structure with thin lignified wall. These cells help in the transportation of water and some nutrients from the roots to the leaves and also provide mechanical support to the plant. Lotus fibers are arranged in these cells in the form of "**Helix**" [21].

Bundle: Fibers taken out from all the cells (holes) are considered as a bundle, as shown in Fig. 3. It is very difficult to extract a continuous uniform length of fiber from one cell (hole). So, in this study a bundle was taken for the following tests: tensile strength, denier and length. For the microscopical study and testing of fiber diameter, single unit was taken.

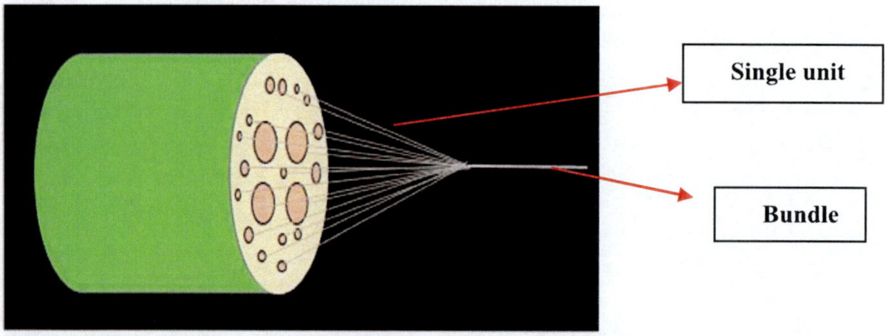

Fig. 3 Arrangement of fibers in the petioles

2.3.1 Identification of the Fiber Through Microscopic, Burning and Solubility Tests: Microscopic Test

Longitudinal view of the fibers was observed under the compound microscope with the magnification of 10× and 45×.

Cross-sectional view of the fiber was tested at Bombay Textile Research Association (BTRA), Ghatkopar Mumbai. For the cross-section, fibers were first coated in JEC gold-plated twin coater, and cross-sectional view was observed using JSM scanning electron microscope with the magnification of 50× to 3500×.

Burning test

To recognize the fiber by burning test, the fiber was moved slowly toward the flame and the reaction of heat was observed. One end of the fiber was put directly into the flame to determine the burning rate and characteristics. The burning odor was noticed.

Solubility test

Fibers were treated with acids and alkalies in both hot and cold conditions.

2.3.2 Material Characterization of the Fiber

Chemical constituents of the raw fibers were determined as per the test suggested by Turner and Doree [8].

2.3.3 Physical Properties of Fiber

Determination of fiber length:

To determine the length of lotus fiber, the fiber was first placed against the dark surface and the length was measured with the steel ruler. An average of 50 readings was taken.

Determination of fiber fineness:

Following the ASTM-D 7025-09 standard [9], fiber fineness was tested. Bundle fiber was taken for testing the fineness. In the direct system of yarn numbering, denier was determined by taking the average weight of 20 readings of 80 cm length of the fiber, and the calculation was done using the formula:

$$\text{Denier} = \frac{\text{W} \times 1}{\text{L}}$$

where W = weight of the fiber.

L = length of the sample

l = unit length of the system

The count of the fibers was also determined by indirect system of yarn numbering using Beasley's yarn balance. The instrument consists of hook at one end and a pointer at the other end. A standard weight (0.040 g) was hung on the notch of the beam. Template was used to cut the length of fibers based on cotton count system. The fibers were kept on the hook until the pointer reaches the datum line. The count is the number of short length filament fibers used to balance the beam.

Determination of fiber diameter:

Compound microscope with micrometer lens was used to measure the fiber diameter. An average of 50 readings was taken to determine the fiber diameter. Single fiber from the petiole was taken to test the diameter.

Determination of fiber strength:

Following the ASTM-D 3822-07 standard [10], fiber strength was tested. LLOYD tensile testing instrument was used. The sample length was 10 cm. The instrument worked on constant rate of elongation principle (CRE). The capacity of the instrument was 2500N. Pulling speed was 100 mm/min. Bundle of the fiber was taken from the petiole for testing the strength.

Moisture testing.

To determine the moisture content and regain, fibers of 10 g were kept in oven for 4 h. After 4 h, sample were taken out from oven and weighted. From the two weight differences, moisture content and regain were calculated using the following formula [11]:

$$\text{Moisture Regain (R)} = \frac{100 \times W}{D}$$

$$\text{Moisture Regain (R)} = \frac{100 \times W}{D + W}.$$

where oven dry weight = D.

$$\begin{aligned} \text{Weight of the water} &= W \\ \text{Regain} &= R \\ \text{Moisture content} &= M. \end{aligned}$$

Fig. 4 Lotus yarn spinning on box (peti) charkha

2.4 *Yarn Spinning*

The extracted lotus fibers were spun on box *(peti) Charkha.* Spinning of yarns was done by Mr. Bakul Shah at Sardar Bhavan Trust, Vadodara. Thus, handspun yarns were prepared.

Extracted fibers were wounded on pern. The spinner first sets the charkha. To start the spinning process, the researcher slowly unwound the extracted fiber from the pern and the spinner hold the fiber from right hand taking the fiber forward toward the point of spindle. The spindle rotated; meanwhile, spinner also rotates the main wheel from left hand in clockwise direction. The twisted yarn was collected on the spindle and later winded on a wooden winder and hanks were prepared as shown in Fig. 4.

2.5 *Yarn Testing*

Based on standard test method ASTM-D 885 [12], yarn testing was done at the testing lab of Century Rayon (under the management and operation of Grasim Industries Limited, Shahad, Mumbai).

2.5.1 Determination of Yarn Fineness

Using direct system of yarn numbering, denier of the yarn was determined by using motorized wrap reel. 90 m of yarn was winded on wrap reel. The hank was taken out from the reel and weighed on digital balance.

The count of the fibers was also determined by indirect system of yarn numbering using Beasley's yarn balance. The instrument consists of hook at one end and a pointer at the other end. A standard weight (0.040 g) was hung on the notch of the beam. Template was used to cut the length of fibers based on cotton count system. The fibers were kept on the hook until the pointer reaches the datum line. The count is the number of short length filament fibers used to balance the beam.

2.5.2 Determination of Shrinkage in Yarns

Shrinkage of the yarns was determined by using wet bulb tube filled with cold water. For the test the yarn of known length was measured. One end of the yarn was knotted with small clip and then dipped in the water and other was knotted in the hook outside the tube and kept for 30 min. After 30 min the yarn was taken out from the tube and kept for drying at room temperature. Then the length was measured and calculations were done using the following formula:

$$\%\,\text{Shrinkage} = \frac{\text{Original length} - \text{Shruken length}}{\text{Original length}} \times 100$$

2.5.3 Determination of Yarn Strength

Tensile strength of the yarns was determined using Instron universal tensile tester. The instrument works on constant rate of elongation principle (CRE). For the test the yarn of known denier was clipped between two jaws with the gauge length 500 mm (50 cm), and the pulling speed was 500 mm/min.

2.5.4 Determination of Yarn Twist

The amount of twist was calculated on Alfred Suter twist tester. The sample length was 10 in. for test length with tension arrangement.

2.5.5 Yarn Wicking

For determining the wicking property of the yarn, the setup was fabricated. The frame of the apparatus was made using stainless steel bar. This bar was supported on the base. A horizontal clamp bar was attached at the top of this bar, which was used to clamp the specimen. A stainless steel scale was attached to the clamp vertically. A glass reservoir was placed on the base of the frame. The reservoir was filled with 250 ml of 1 % dye solution. Two clips were used, one for clamping the sample at the top and other at the bottom of sample to keep it in a straight position.

The transport of liquids in a porous media is driven by capillary forces that arise from the wetting of yarns. The rise of liquid along the specimen was recorded on the basis of the height shown against the scale. At the same time, a stop watch was also started in order to record the time it took the moisture to travel from mark to mark. This was done in order to compare length traveled by liquid versus time. The rise in length was recorded at different intervals of time. For testing of yarn, a length of 180 mm was clamped at the top ensuring immersion of 30 mm in the liquid [13].

2.6 *Construction of Fabric*

Union fabric sample was constructed using 100 % lotus handspun yarn as a weft and cotton as a warp. To compare the properties, fabric with cotton (warp and weft) and silk (warp and weft) were also prepared. Weaving was done in treadle loom at Bhujodi, Kutch.

2.7 *Evaluation of the Properties of Constructed Fabric*

2.7.1 Determination of Fabric Count

Fabric count (the number of yarns/inch) helped to describe the closeness of the weave. Following the ASTM-D 3775-98 [14] standard fabric count was determined by counting the number of threads in 1 in. in warp and weft direction using pick glass.

2.7.2 Determination of Fabric Thickness

Following the ASTM-D 1777-96 [15] standard fabric thickness was measured using universal thickness tester.

2.7.3 Determination of Fabric Weight Per Unit Area:

Following the ASTM-D 3776 [22], sample of 5 × 5 cm was cut and weighted. GSM was calculated using the following formula:

$$\text{GSM} = \frac{\text{Weight in grams of the sample} \times 100 \times 100}{5 \times 5}$$

2.7.4 Determination of Tensile Strength of the Fabric

The tensile strength of the fabrics was tested on universal tensile tester (UTM) by raveled strip test method using ASTM-D 5035-95 [16]. Specimen size of 150 mm× 25 mm (15 cm× 2.5 cm) was cut. Gauge length was kept 75 mm ± 1 mm (7.5 cm) with the speed of 300 mm/min.

3 Results and Discussion

3.1 Fiber Extraction

The portion between the flower and rhizome of the lotus plant is called "petiole" which was used for fiber extraction. Being a natural plant, to obtain the consistency in the petiole diameter, length, color was very difficult. All these parameters vary in different seasons, because the quality of water varies from pond to pond and the maturation of the plant. After collecting the petioles from the pond, the fibers can be extracted till 7 days because day-by-day the petiole gets dry and the continuous length of the fibers cannot be obtained for uniform yarn preparation. The extraction process is mentioned in the following steps:

Step 1: The first and foremost step was to wash the lotus petiole in running tap water to remove all kinds of dirt from the outer body of the petiole (Fig. 5a).

Step 2: After washing, petioles were wiped with fresh cloth (Fig. 5b).

Step 3: The bunch of three petioles was laid on wooden slab (length and width of the wooden slab was 58 and 28 cm, respectively) and was slit 5–7 times at different sections as per the length of the petiole with the help of a sharp knife (Fig. 5c).

Step 4: After one slit, a bunch of three petioles was slowly stretched apart to extract the fibers. Fibers are naturally present inside the petiole in two xylem cells: tracheids and vessels in the form of "helix". So as we stretch the petioles, the fibers come out from each cell, and during the time of extraction little amount of water also comes out from the petiole because it is a water plant. The extracted fibers were laid on the wooden slab (Fig. 5d).

Step 5: Initially the extracted fibers were spun by palm twisting, which is a very time-consuming process. So, later fibers were directly taken on the pern as shown in Fig. 5. Only the end-to-end points were joined by palm twisting (Fig. 5e).

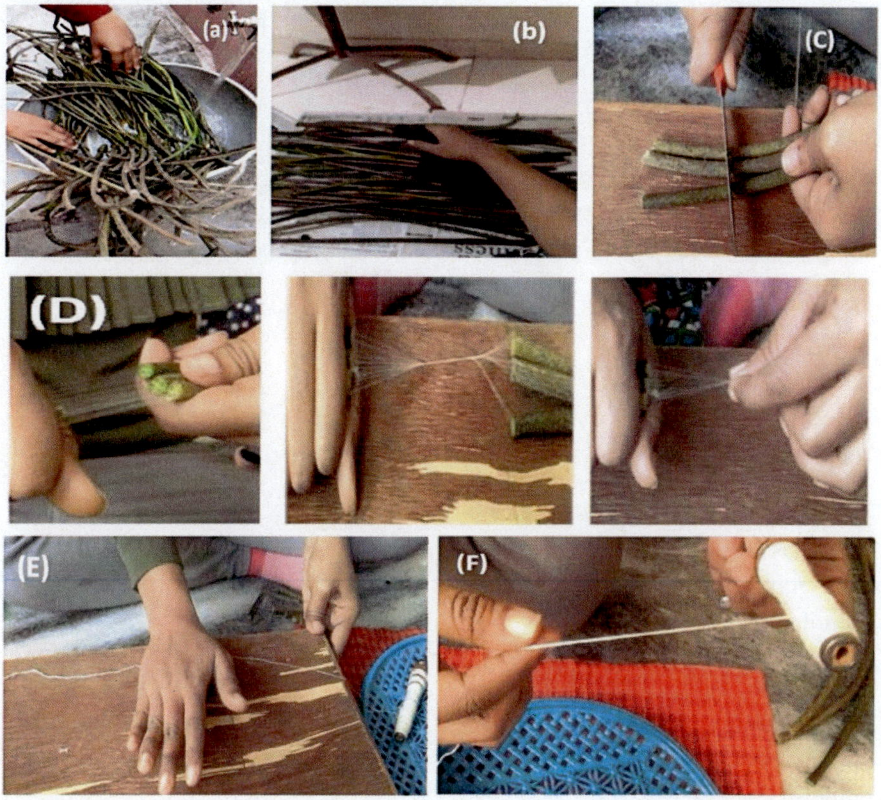

Fig. 5 Extraction of lotus fibers: **a** Washing of petioles; **b** washed petiole wiped with fresh cloth; **c** bunch of three petioles laid on wooden slab; **d** stretching of fibers while twisting; **e** end-to-end points joined by palm twisting; **f** extracted fiber wounded on pern

4 Fiber Testing

4.1 Identification of Fiber by Microscopic Appearance, Burning and Solubility

Microscopic appearance.

Longitudinal view: It was observed that single unit was smooth and fine under 10× magnification, as shown in Fig. 6a, and the bundle has a wavy structure. Several individual fibers are conglutinated together into one bundle under 5×, 10× and 45× , as shown in Fig. 6b.

Cross-sectional view: The cross-section of the lotus fiber was observed by scanning electron microscope in magnification 50×, 100×, 150×, 200×, 300×, 400×,

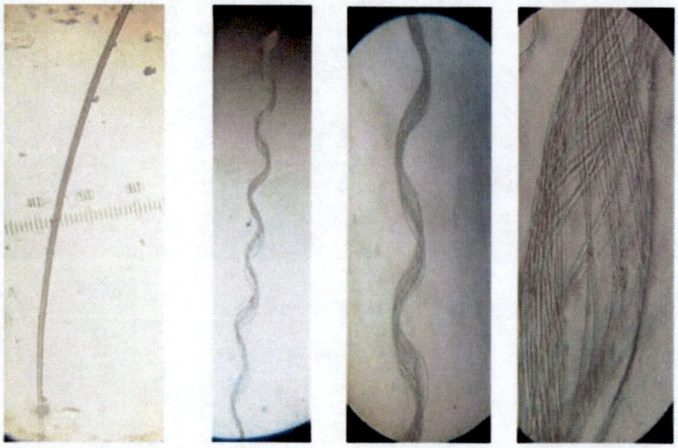

a b

Fig. 6 Longitudinal view of lotus fiber **a** Single unit (10×); **b** Bundle (5×, 10×, 45×)

500×, 2000× and 3500×, as shown in Fig. 7. It was observed that fiber has circular or similar to circular shape.

Solubility test (Table 1).

Lotus fibers were subjected to various acids and alkalies in both hot and cold conditions. It was found that fibers were dissolved in 99 % sulfuric acid in cold condition. Hence burning and solubility test confirm and exhibit the characteristic of cellulosic fiber.

4.2 Material Characterization of Fiber

The chemical composition of the fiber obtained by the elemental analysis was done as per the test suggested by Turner and Doree. The values are presented in Fig. 8.

4.3 Physical Properties of Fiber

Fiber Length:

Length of the lotus fiber was recorded between 60 and 105 cm and the average length was 80.70 cm. Fibers shorter than 15 mm tend to have insufficient length and are to be twisted into yarn structure; the fibers longer than 150 mm tend to require specialized spinning machinery for the conversion into yarn structure. Hence it must have enough length to be twisted and converted into yarn structure [17].

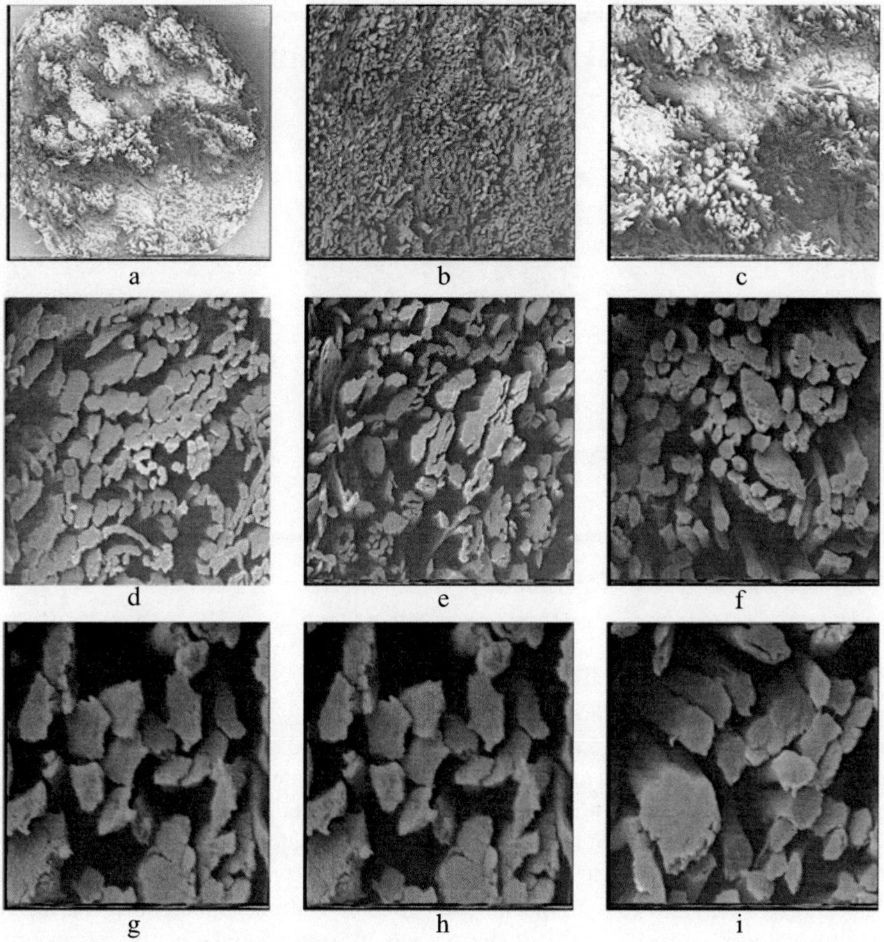

Fig. 7 SEM images of lotus fiber (cross-sectional view) **a** 50×, **b** 100×, **c** 150×, **d** 200×, **e** 300×, **f** 500×, **g** 1000×, **h** 2000×, **i** 3500×

Table 1 Burning test of lotus fiber

Fiber type	Cellulosic fiber
When approaching flame	Catches fire rapidly
When in flame	Burn with light grey smoke
After removal from flame	Continues to burn afterglow
Residue	Soft black crushable ash
Odor	Burning paper

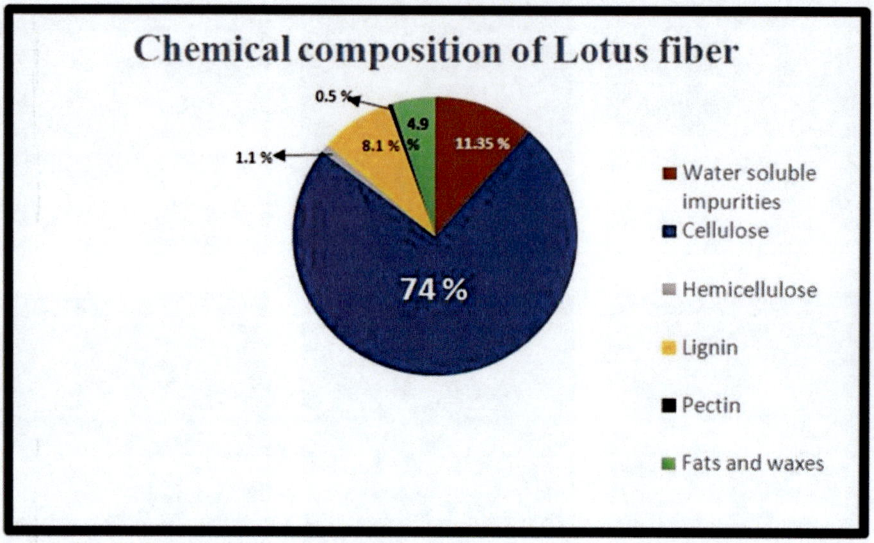

Fig. 8 Chemical composition of lotus fiber

Fiber Diameter:

For fiber diameter only a single fiber from the petiole was tested. It was observed that diameter was between 2 and 6 μm and the average diameter was 3.5 μm. Fibers or filaments finer than 10 μm tend to become too delicate for ready conversion into yarn structure [17]. Hence the diameter of the single lotus fiber was much smaller than other plant fibers and it is less than 5 μm, which belongs to the category of the microfiber. Unlike the synthetic superfine microfiber, it is natural and easily obtained without any use of technology and chemical process. A microfiber textile has a soft feeling and large surface area with an excellent absorption and adsorption capacity [19].

Length to breadth ratio:

An average length of the lotus fibers observed was 80.70 cm, and the average diameter observed was 3.5 μm, leading to length to breadth ratio of 269,000:1. The fact associated with this ratio is higher the ratio, finer is the fiber, and lower the ratio, coarser is the fiber. The smallest suitable ratio of fiber length to fiber breadth (thickness) is about 350:1. Anything less than this indicates a fiber that probably will not permit twisting into a yarn structure. But a ratio of 1000:1 or more indicates a fiber that should readily be spun into a yarn structure [17]. Hence the high length to breadth ratio of lotus fiber indicates that it can readily be spun into a yarn.

Table 2 Fineness of lotus fiber

Fiber	Denier	Tex	Cotton count (Ne) (Indirect system)
Lotus Fiber	32	3.5	166 s

Table 3 Fiber strength

Fiber sample	Maximum load (gf)	Extension at maximum (mm)	Stress g/den	Strain (%)
Lotus fiber	161.52	1.5589	5.0476	1.5589

Fiber fineness:

In direct system, denier was calculated and the denier obtained was converted into Tex. Count was also calculated by indirect system as per cotton count. The values obtained are given in Table 2:

In the direct system, the lower the denier, the finer is the fiber, and the higher the denier, coarser is the fiber and as per indirect system, the lesser the count, coarser is the fiber, and the higher the count, finer is the fiber. Hence lotus fiber is fine in character.

Fiber strength:

Load and elongation, stress and % strain values of lotus fiber were tabulated in Table 3. It was observed that lotus fiber has poor elongation property because all the minor fibers have a low extension value and any kind of finishing can improve the strength.

Fiber Moisture:

Moisture affects the physical properties of the fiber: dimensional property, mechanical properties including breaking strength, elongation, crease recovery and electrical properties. Amount of moisture depends on relative humidity and temperature. Fiber strength is greatly influenced by moisture regain. Moisture content of the lotus fiber was 10.6 % and regain was 11.8 %, which was more than cotton and close to silk and viscose rayon fiber.

4.4 Yarn Spinning

It is twisting together of drawn-out strands of fibers to form yarn. For better spinnability, the fiber must have better cohesiveness means and they must hold together to prevent slippage [18]. The extracted fibers were spun into yarn on box (*peti*) *charkha*/foldable spinning wheel. It is a manual hand-driven charkha used for spinning cotton fibers.

Table 4 Fineness of 100 % lotus handspun yarn

Yarn sample	Denier	Tex	Cotton count (Ne)
Lotus yarn	108	12	50 s

Table 5 Shrinkage (%) of lotus yarn

Initial Length (in cm)	After (in cm)	Difference (in cm)	Shrinkage (%)
54.5	54.5	0	0
67.4	67.4	0	0
60.3	60.3	0	0
65.5	65.5	0	0
67.2	67.2	0	0

Initially, the fiber bundle with one and two petioles was experimented for spinning. But it was not easily spinnable and there were more breakages. But the fiber bundle taken from three petioles was easily spinnable. For entire study the fiber bundle from three petioles was spun. During spinning it was observed that fibers were spinnable without fewer breakages in entire spinning process.

4.5 *Yarn Testing*

4.5.1 Yarn Fineness

Fineness of the yarn is important to obtain a soft, smooth and uniform fabric [18]. The fineness of the yarn was determined with both direct and indirect system. The values obtained have been given in Table 4.

4.5.2 Yarn Shrinkage

Shrinkage is the dimensional change in the reduction of length in yarn. The values of the shrinkage (%) have been given in Table 5. Hence it was observed that lotus yarn has 0 % shrinkage, which is an advantage over the other natural fibers like cotton and wool where there is a problem of shrinkage.

4.5.3 Yarn Twist

The spiral disposition of the components of a thread is usually the result of relative rotation of the two ends [11]. Twist per inch (tpi), twist per meter (tpm) and twist

Table 6 Twist tpi/tpm of lotus yarn

Yarn sample	Twist per inch (tpi)	Twist per meter (tpm)	Twist direction
Lotus yarn	1.66	65.6	S

Table 7 Tensile strength of lotus yarn

Yarn Sample	Maximum load (gf)	Extension at maximum, mm	Stress, g/den	Strain (%)
Lotus (handspun)	288	1.7	2.66	1.7

direction of the handspun lotus yarn are listed in Table 6. It was observed that lotus yarn was a low twist of 1.66 tpi.

4.5.4 Yarn Strength

Strength and elongation is an important property of any yarn to be used for weaving. Load and elongation, stress and strain values of the lotus handspun yarn have been given in Table 7. It was observed that lotus yarn has low elongation property.

4.5.5 Yarn Wicking

Wicking is a spontaneous transport of a liquid driven into a porous system by capillary forces. The rise of the liquid along the yarn was noted versus time in minutes and is tabulated in Table 8. Hence it was observed that the rate of rise of liquid in the yarn was much higher for first few minutes and then it decreased gradually. So, the fabric constructed from lotus yarn will have a good comfort property.

4.6 Fabric Construction and Its End Uses

During the fabric construction, yarn fineness of both warp and weft was kept in mind. The weave type was kept constant for all the fabric and that was plain weave. Union fabric samples were prepared from 100 % lotus handspun yarn used as a weft and cotton as a warp, as shown in Fig. 9, to compare the properties of fabrics with cotton (warp and weft) which was also prepared.

The specifications of the yarns used for the construction of the fabrics on handloom are given in Table 9.

Table 8 Wicking of lotus yarn

	Height (in cm) rate of rise of liquid in the yarn				
Time (min)	Sample-1	Sample-2	Sample-3	Sample-4	Sample-5
0.5	8.2	9	10	9.8	9.4
1	9	10	10.8	10.2	10
1.5	11	12	12.2	10.8	10.8
2	12	13.2	12.6	11.2	11
2.5	12.5	13.5	12.8	12.5	11.5
3	12.8	13.8	13	12.8	12.6
3.5	13	14	13	13	13
4	13	14	13	13	13
4.5	13	14	13	13	13
	13	14	13	13	13
	Height (in cm) rate of rise of liquid in the yarn				
Time (min)	Sample-6	Sample-7	Sample-8	Sample-9	Sample-10
0.5	9	10	9.5	9.7	9.3
1	9.5	10.7	10.6	10.4	9.8
1.5	10.3	11	11	11.3	10.2
2	11	11.2	11.5	11.8	10.9
2.5	12.5	11.5	11.8	12	11.5
3	12.7	11.8	12	12.2	12
3.5	13	12	12	12.4	12
4	13	12	12	12.4	12
4.5	13	12	12	12.4	12
5	13	12	12	12.4	12

a b

Fig. 9 Construction of fabric samples on handloom at Bhujodi, Kutch. **a** Preparation of fabric on treadle loom, **b** Cotton: Lotus fabric

Table 9 Yarn specifications for the construction of fabric

S. no	Fabric sample	Warp yarn		Weft yarn	
		Fiber content	Yarn count	Fiber content	Yarn count
1	100 % Cotton	Cotton	2/80 s	Cotton	2/80 s
2	Cotton: Lotus	Cotton	2/80 s	Lotus	50 s

Table 10 Fabric count of constructed fabrics

S. no	Fabric sample	EPI (ends per inch)	PPI (picks per inch)	Fabric count
1	100 % Cotton	50	42	50 × 42
2	Cotton: Lotus	50	44	50 × 44

4.7 Evaluation of the Properties of Constructed Fabric

4.7.1 Determination of Fabric Count

Fabric count is counting the number of warp and weft threads in 1 in. in both warp and weft direction using pick glass. The count of fabric samples is mentioned in Table 10. It was observed that there was not much difference in readings between 100 % cotton and cotton:lotus fabric.

4.7.2 Determination of Fabric Thickness

Thickness of the samples is mentioned in Table 11. It was observed that there was no major difference in thickness between the cotton and cotton:lotus fabric sample.

Table 11 Thickness of constructed fabrics

S. no	Fabric sample	Thickness (mm)
1	Cotton	0.28
2	Cotton:Lotus	0.29

Table 12 GSM of constructed fabrics

S. no	Fabric sample	GSM (in grams)	GSM (in ounces)
1	Cotton	70.8	2.48
2	Cotton:lotus	80.8	2.85

Table 13 Tensile strength of constructed fabrics

S. no	Fabric sample	Max load (kgf)	Max extension (mm)	Strain (%)
1	100 % Cotton (warp)	6.45	6.41	8.52
2	100 % Cotton (weft)	3.69	3.21	4.28
3	Cotton:Lotus (warp)	7.75	6.79	9.05
4	Cotton:Lotus (weft)	7.49	4.6	6.13

4.7.3 Determination of Fabric Weight Per Unit Area

GSM of fabric samples is mentioned in Table 12. It was observed that GSM of cotton:lotus fabric was more than 100 % cotton fabric. GSM between 1 and 4 oz falls under lightweight category [20]. Thus, the GSM of fabric samples prepared fall under the category of lightweight fabrics.

4.7.4 Determination of Tensile Strength of the Fabric

The results of load and elongation give very important information about the tensile behavior of fiber, yarn and fabric. Load and extension values of two fabric samples are mentioned in Table 13. It was observed that load and extension of cotton:lotus fabric was higher than 100 % cotton fabric.

5 Conclusion

Moving to the eco-friendly environment, peoples across the globe are more forward toward the use of natural fibers due to its environmental, health and economic benefits over synthetics. Natural fibers like cotton, silk, wool and jute have been explored and currently used by peoples. But there are many other natural minor fibers like ramie, sisal, banana, coir, water hyacinth, raffia, abaca, hemp, nettles which are coming up in various textile applications. These minor fibers are extracted from the part of the plant that goes as an agricultural waste which has a potential to be used as a textile fiber. This study was done to explore the lotus petiole (that is the part between flower and rhizome) of the lotus plant that goes as waste after cutting the flower. The petiole contains precious fiber that has a potential to be textile fiber. The researcher started the study with the search of lotus wetlands and contacting the peoples involved in cultivation and marketing of the lotus to know the availability of the waste in bulk. Petioles were procured from cultivators and fibers were extracted. The extracted fibers were tested for its microscopic appearance, burning, solubility, chemical composition, fiber length, fiber diameter, length to breadth ratio, strength, fineness moisture content and regain. The extracted fibers were experimenting for its spinnability on peti charkha and the yarn was also subjected to following test: yarn fineness, strength, shrinkage, twist, moisture and wicking. Union fabric samples were prepared using 100 % lotus handspun yarn as weft and cotton as a warp. To compare the properties, fabrics with cotton (warp and weft) was also prepared. The constructed fabric was also subjected to the following test: fabric count, GSM, thickness and strength. Except human energy no other form of energy was used in fiber extraction. Being manual in nature, the process is time-consuming. The extracted fiber was easy spinnable on *peti charkha* without less yarn breakages in entire spinning process. 100 % lotus handspun yarn was prepared. It was found that load and extension value of cotton: Lotus fabric was higher than 100 % cotton fabric. Results suggest that lotus fiber can be used in yarn and fabric stage for preparing union and blended fabrics with cotton, silk and viscose. As a waste is abundant raw material even for its use on larger scale will not be a problem. The process from extraction to weaving will not require use of any chemical which will be an added advantage to green environment.

Acknowledgements The authors are thankful to Mr. Poonambhai Parmar, Mr. Sanjaybhai Mali, Mr. Isabbhai Rathod (Cultivators of Lotus), Mrs. Shantaben Mali (Vendor of lotus flower), and Mr. Bakul Shah (Member of Sardar Bhavan Trust) of Vadodara district. Mr. Himanshu Agarwal (Deputy General Manager), Mr. Nishith K. Sheth (Asst. Manager Tyrecord and CSY plant) of Century Rayon (under the management and operation of GRASIM Industries Limited), Shahad, Mumbai and Bombay Textile Research Association (BTRA), Ghatkopar, Mumbai for helping me in various stages of research.

References

1. Gardetti M, Muthu S (eds) (2015) Handbook of sustainable luxury textiles and fashion, vol 1. Springer, Singapore, pp 59–98
2. Saha GK, Mazumdar S (2017) Wildlife biology: an Indian perspective. PHI Publishers, New Delhi
3. https://bsienvis.nic.in/files/National%20Flower_Nelumbo%20nucifera_26.9.14.pdf . Accessed 01 Feb 2019
4. Myint T, San D, Phyo U (2019) Lotus fiber a value chain in Myanmar. Helvatas:1–40. Retrieved from themimu.info>node
5. Sheikh SA (2014) Ethno-medicinal uses and pharmacological activities of lotus (Nelumbo Nucifera). J Med Plants Stud 2(6):42–46. https://www.researchgate.net/publication/293183331
6. https://biologydictionary.net/petiole/. Accessed 20 Mar 2019
7. https://biologydictionary.net/rhizome/. Accessed 20 Mar 2019
8. Garner W (1967) Textile lab manual. 5(3)
9. ASTM D 7025-09 (2015) e1: Standard Test method for assessing clean flax fiber fineness. ASTM International, West Conshohocken, PA
10. ASTM D3822-07 (2007) Standard test method for tensile properties of single textile fibers. ASTM International, West Conshohocken, PA
11. Booth JE (1996) Principles of textile testing. CBS Publishers and distributors Pvt. Ltd. New Delhi, India, pp 100–230
12. ASTM- 885/D885-10 (2014) A e^1 Standard test methods for tire cords, tire cord fabric, and Industrial filament yarns made from Manufactured Organic base fabrics. ASTM International, West Conshohocken, PA

13. Roy MD, Sinha SK (2014) Performance of wicking through yarn and fabric made from polyester fibres of different cross- sections. Int J Text Sci 3(3):44–50. https://doi.org/10.5923/j.textile.20140303.02
14. ASTM D 3775-98 (1998) Standard test method for fabric count of woven fabric. ASTM International, West Conshohocken, PA
15. ASTM D 1777-96 (2019) Standard test method for thickness of textile materials. ASTM International, West Conshohocken, PA
16. ASTM D 5035-95 (1995) Standard test method for breaking force and elongation of textile fabrics (Strip Method), ASTM International, West Conshohocken, PA
17. Gohl EPG, Vilensky LDT (1987) CBS Publishers & Distributors Pvt Ltd. New Delhi India
18. Mishra SP (2018) A textbook of fibre science and technology. New Age International Publishers London, pp 4–5
19. Zhao L, Chen DS, Gan YJ, Yuan X, Wang Y (2015) Analysis of length and fineness of lotus fiber extracted by physical methods. Chem Eng Trans 46:85–90. https://doi.org/10.3303/CET1546015
20. Young D (2014) Swatch reference guide for fashion fabrics. Bloomsburg Publications New York
21. Mengxi W, Hua S, Qunfeng C, Lei J (2014) Bioinspired green composite lotus fibers. Angewandte Commun 53:3358–3361. doi:https://doi.org/10.1002/anie.201310656
22. ASTM D3776/D3776M-09a (2017) Standard test method for Mass Per Unit Area (Weight) of Fabric, ASTM International, West Conshohocken, PA

Green Manufacturing Model for Indian Apparel Industry Using Interpretive Structural Modeling

Ankur Saxena and Ajit Kumar Khare

1 Introduction and Literature Review

1.1 Green Manufacturing and Its Advantage

Researchers have given multiple definitions of green manufacturing; one of the definitions given by Das et al. states that there are manufacturing methods that support and sustain a renewable way of producing products and services that do no harm to you or the environment [1, 2]. Green manufacturing is defined as the design, processing and commercial use of materials processes and products, which are economical and sustainable while minimizing pollution and risk to human health and the environment. Green manufacturing is more of a philosophy rather than an adopted process or standard [6].

Green manufacturing is a need for sustainable development and a means of competitive advantage for the firms [7]. It is considered that firms which successfully implement green manufacturing are preferred by customers. These manufacturing organizations continuously strive hard to innovate strategies to adopt the green practices to reorganize the supply chain structures, so that benefit of strategic green manufacturing practices can be achieved in the product management [8–11]. The value of investing in green technology and green transformation is considered as a

A. Saxena (✉)
National Institute of Fashion Technology, Jodhpur 342017, India
e-mail: ankur.saxena@nift.ac.in

A. K. Khare
National Institute of Fashion Technology, Mumbai 410210, India

A. Majumdar et al. (eds.), *Functional Textiles and Clothing 2020*,
https://doi.org/10.1007/978-981-15-9376-5_14

topic of argument between researchers and managers [12]. Hoffman states that environmental and green attempts in manufacturing should move from being an environmental management approach to an environmental strategy. This will create a win-win situation by which manufacturers can improve their environmental performance while achieving economic gains [13].

1.2 Green Manufacturing Practice in Apparel Industry

Although literatures are available on the use of green manufacturing or reducing carbon footprint in different industries around the globe [14–16], researchers talk about green manufacturing in several industries like cement, automobiles and so on [17]. In the context of global apparel industry, Eryuruk et al. calculated carbon emission during a lifecycle of an apparel product. He divided the life cycle from design to reuse and calculated emission for each process [18]. Guo-Ciang et al. presented a model for Taiwan textile and apparel industry in which drivers for green supply chain management were identified [19]. Several researchers have discussed about evaluating present green manufacturing practices in different department of apparel and textiles globally [20, 21].

1.3 Global Warming and Its Effect on Climate

In 2016, greenhouse gases emission was 58 giga tonnes (Gt.) globally, which include 196 countries, and around 5 % of total countries are responsible for more than 85 % of the total emission. Industrialization, energy and agriculture are the major factors contributing more than 80 % of the total global emission. In the global context, India is the fourth largest emitter of carbon dioxide at 1.65 Gt. per year, after China (6.9 Gt. per year), United States (5.2 Gt. per year) and European Union (2.5 Gt. Per year) [3]. The effects of climate change, along with pollution and the depletion of non-renewable natural resources, have given rise to environmental awareness [1]. Since the early twentie century, the average surface temperature of the earth has increased by about 0.8 °C (1.4 °F), with about two-thirds of the increase occurring since 1980. Researchers indicate that during the twenty-fir century, the global surface temperature is likely to rise by a further 1.1–2.9 °C (2–5.2 °F) for the lowest emissions scenario and 2.4–6.4 °C (4.3–11.5 °F) for the highest emission scenario. Warming of the climate system is irreversible, and scientists are in the view that most of it are caused by increasing concentrations of greenhouse gases produced by activities such as deforestation and the burning of fossil fuels [2]. As per the report of World Resourses Institute, India is producing over 7 % of global greenhouse gases. Industries, agriculture and power generation are the main factors that produce more than 90 % of Indian emission [4]. According to Mehta et al., the textile industry affects the environment by releasing carbon dioxide into the atmosphere. The weaving and spinning sector contributes a lot to emission in carbon dioxide. However, the apparel sector will contribute 3–4 % of overall emission and is increasing with a rapid growth [5]. To reduce the greenhouse gas emission and further to minimize the adverse effect on the environment, the importance and effect of green manufacturing have created a much wider debate in recent years.

1.4 What is a Framework?

The dictionary meaning of a framework is:

"A set of assumptions, concepts, values, and practices that constitute a way of viewing reality" [2], while in the context of the research in the garment industry it is defined as "It is a tool that can be used by any garment industry irrespective of size or product to access present status in term of green manufacturing, further, it will also be helpful to the organization to improve its functionality of green manufacturing towards positive side". The framework is a basic structure underlying a system, concept or text [3].

Miles and Huberman (1994) defined a conceptual framework as a visual or written product, one that "explains, either graphically or in narrative form, the main things to be studied—the key factors, concepts, or variables—and the presumed relationships among them" [5].

2 Need of the Research

As per Prof. Mukesh Sharma, the annual per capita tCO2e emission in the city of Delhi was estimated as 2.26 ton in 2016, which is 1.5 times of national average [24]. Also, as per the research by Dr Deveraja under Indian Council of Social Science-Ministry of HRD, Government of India, Delhi region is considered a significant cluster of apparel manufacturing in India and contributes 8.7 % in export and 10.9 % in domestic market [25]. Statement of TV Ramachandra and Ministry of Textiles (MOT) report confirms that there is a problem of CHG emission in Delhi/NCR. Industrialization in general and apparel industry of the region in particular are the major causes for the same. A significant fraction of the apparel industry is in Delhi/NCR which is also the highest emitter of GHG because of industrialization. As per a book written by TV Ramachandra et al. "Sector-Wise Assessment of Carbon Footprint across Major Cities in India", eight cities including all metros were studied and their sector-wise greenhouse gases emissions were calculated. Industrial CHG emission per capita was highest in Delhi NCR during 2014–2015 [23].

Garment industry being a defragmented and unorganized industry requires a systematic framework for green manufacturing which will help manufacturers to analyze their current status and further help to take suitable policy decisions to make their process more sustainable and green.

3 Methodology and Research Design

The review of existing literature on development of frameworks for green manufacturing for different industries led to the selection of following tools:

1. Fuzzy Delphi method (FDM)
2. Interpretive structural modeling (ISM)

Figure 1 shows the process flow followed to achieve the objective.

3.1 *Sample Size*

It is established in "Journal of industrial statistics" in 2015 about efficiencies and policies of Indian apparel industry that a high scale of operations helps the adoption of modern technology in Indian apparel industry, and the size of the firm may significantly decide the efficiency and policy making of the firm in a positive manner. The researcher has also divided the garment factories into four quadrants according to their sizes and policies [26]. Learning from this, it was decided to include the factories which are in quadrant three and four, that is, factories having more than 400 sewing machines. There are around 800 such factories in Delhi/NCR (Source: Apparel Resources yellow pages). It was suggested by the research committee to have minimum 10 % (80 factories) responses in order to obtain valid opinion. To ensure minimum required responses, around 200 respondents of different apparel manufacturing units using purposive sampling were approached, out of which 81 responses were found complete and included for analysis using SPSS, a statistical tool.

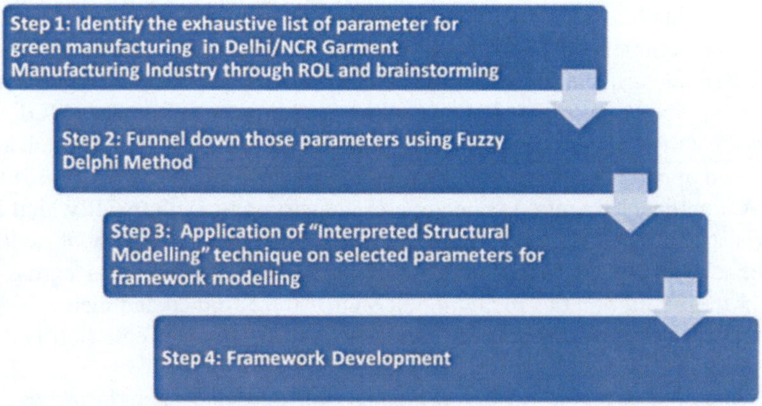

Fig. 1 Methodology process flow

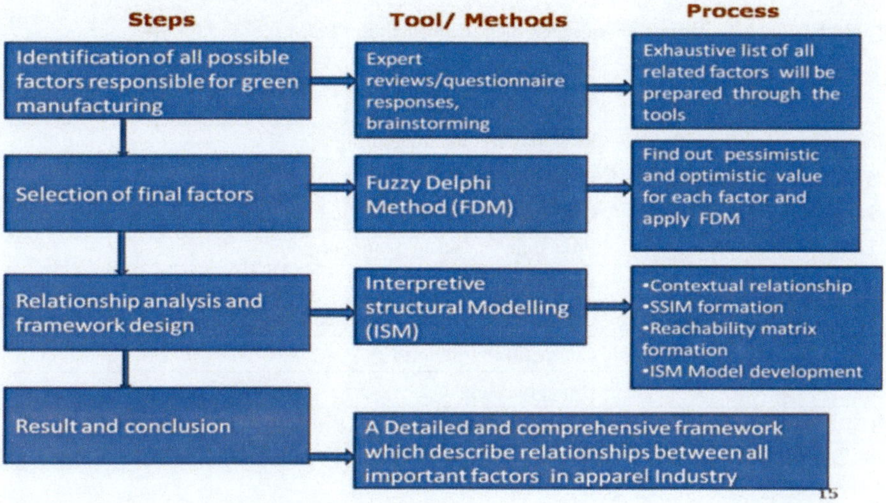

Fig. 2 Methodology of framework development

Data Collection and Analysis

India is a significant and large emitter of greenhouse gases and most of it because of the industrial production. Hence there is a certain need to reduce these emissions, justifying thereby the focus is on investigating the awareness and attitude toward green manufacturing in the apparel industry of Delhi/NCR. A panel of five experts was identified and based on their reviews and opinion FDM and ISM were used to develop a green manufacturing framework.

3.2 *Process Flow of Framework Development*

The review of existing research on the subject led to the identification of following process flow to be adopted for green manufacturing framework for apparel industry. Figure 2 explains the steps used in framework development.

3.3 *Use of Interpretive Structural Modeling*

To design green manufacturing framework, an exhaustive list of 19 parameters was prepared through expert reviews and brainstorming. These parameters were subsequently funneled down through fuzzy Delphi method. Funneling process of FDM led to the reduction of the following 10 parameters (Table 1), which were further used while designing green manufacturing framework.

Table 1 Funneled parameters of green manufacturing after FDM

S. No	Code	Parameter
1	F-1	Type of energy used
2	F-2	Raw material manufacturing
3	F-3	Apparel manufacturing technology
4	F-4	Government norms for the industry or regulatory framework
5	F-5	Procedure for waste treatment
6	F-6	Economic constraints
7	F-7	Green logistics (packaging and transportation)
8	F-8	Competitive strategies
9	F-9	Brand building
10	F-10	Location of the factory

The following stepwise procedure is adopted while applying "Interpreted Structural Modeling":

Step 1: Parameters were funneled using FDM technique and listed for next step.

Step 2: Contextual relationships are established among the parameters listed in Step 1.

Step 3: Structural Self-interaction Matrix (SSIM).

- Keeping in view the contextual relationships of each parameter, the existence of an interaction and its direction between any two parameters (*i* and *j*) was enquired from a group of experts.
- Four symbols were used to represent possible interaction that exists between any two parameters i and j.

V: Parameter i influences parameter j
A: Parameter j influences parameter i
X: Both i and j influence each other
O: No interaction between any two parameters

Step 4: Reachability Matrix.

The SSIM format prepared in Step 3 was transformed into the reachability matrix by replacing symbols (V, A, X and 0) with binary digits, 0 and 1. Rule of replacement is as follows (Table 2):

Step 5: Partitions on the reachability matrix.

After the grounding of reachability matrix, it was processed to extract the digraph and associate structural models.

Table 2 Rule of replacement from SSIM to reachability matrix

If the (i, j) entry in the SSIM is	Entry in the initial reachability matrix	
	(i, j)	(j, i)
V	1	0
A	0	1
X	1	1
O	0	0

Step 6: Lower-triangular format reachability matrix.

The reachability matrix was further transformed into a lower triangular format by identifying the highest-level elements and inserting them as the first elements in the new reachability matrix.

Step 7: Digraph for interpretive structural model.

Having identified the levels of the elements, the relationship between the elements was drawn, indicating the serial number of the elements and the direction of relation with the help of an arrow.

Step 8: Interpretive Structural Model.

3.4 Elements, Contextual Relationship and Interpretation

Contextual interactions between all parameters were established with the help of expert's opinion and brainstorming. A total of 90 relations were established between 10 parameters and a separate table is prepared for each. Table 4 is the table describing the contextual relationship between all the parameters (Table 3).

Similar table was formed for each parameter which comprises 10 different tables of 90 relationship.

3.5 Development of SSIM Matrix

Consultation and discussions with lean practitioners, experts and brainstorming sessions helped in identifying the relationships between the identified parameters. With the help of the contextual relationship between parameters, the SSIM has been developed. As mentioned in Step 3, four symbols (V, A, X and O) were used. Table 4 presents SSIM matrix.

Table 3 Contextual relationship of F1 with all other parameters-interpretive logic

No	Paired element	Description	Relation (Y/N)
1	F1–F2	Type of energy used will influence/enhance raw material manufacturing	Y
2	F2–F1	Raw material manufacturing will influence/enhance type of energy used	Y
3	F1–F3	Type of energy used will influence/enhance apparel manufacturing technology	N
4	F3–F1	Apparel manufacturing technology will influence/enhance type of energy used	Y
5	F1–F4	Type of energy used will influence/enhance government norms for the industry or regulatory framework	N
6	F4–F1	Government norms for the industry or regulatory framework will influence/enhance type of energy used	Y
7	F1–F5	Type of energy used will influence/enhance procedure for waste treatment	Y
8	F5–F1	Procedure for waste treatment will influence/enhance type of energy used	Y
9	F1–F6	Type of energy used will influence/enhance economic constraints	N
10	F6–F1	Economic constraints will influence/enhance type of energy	Y
11	F1–F7	Type of energy used will influence/enhance green logistics (packaging and transportation)	N
12	F7–F1	Green logistics (packaging and transportation) will influence/enhance type of energy used	Y
13	F1–F8	Type of energy used will influence/enhance competitive strategies	Y
14	F8–F1	Competitive strategies will influence/enhance type of energy	Y
15	F1–F9	Type of energy used will influence/enhance brand building	Y
16	F9–F1	Brand building will influence/enhance type of energy	N
17	F1–F10	Type of energy used will influence/enhance location of the factory	Y
18	F10–F1	Location of the factory used will influence/enhance type of energy used	Y

3.6 Development of Reachability Matrix

As explained in Step 4, the reachability matrix was prepared from SSIM by transforming the information of each cell of SSIM into binary digits (i.e., 1 s or 0 s). Following these rules, the reachability matrix is prepared. Table 5 presents the reachability matrix.

Table 4 SSIM Matrix

S. no	Parameters/element	10	9	8	7	6	5	4	3	2	1
1	Type of energy used	X	V	X	A	A	X	A	A	X	
2	Raw material manufacturing	A	X	X	X	A	V	A	0		
3	Apparel manufacturing technology	A	X	X	0	A	V	A			
4	Government norms for the industry or regulatory framework	X	0	V	V	V	V				
5	Procedure for waste treatment	X	V	V	V	A					
6	Economic constraints	0	V	V	V						
7	Green logistics (packaging and transportation)	A	V	V							
8	Competitive strategies	A	X								
9	Brand building	A									
10	Location of the factory										

Table 5 Reachability matrix

S. No	Parameters/element	10	9	8	7	6	5	4	3	2	1	Driver
1	Type of energy used	1	1	1	0	0	1	0	0	1	1	**6**
2	Raw material manufacturing	0	1	1	1	0	1	0	0	1	1	**6**
3	Apparel manufacturing technology	0	1	1	0	0	1	0	1	0	1	**5**
4	Government norms for the industry or regulatory framework	1	0	1	1	1	1	1	1	1	1	**9**
5	Procedure for waste treatment	1	1	1	1	0	1	0	0	0	1	**6**
6	Economic constraints	0	1	1	1	1	1	0	1	1	1	**8**
7	Green logistics (packaging and transportation)	0	1	1	1	0	0	0	0	1	1	**5**
8	Competitive strategies	0	1	1	0	0	0	0	1	1	1	**5**
9	Brand building	0	1	1	0	1	0	0	1	1	0	**5**
10	Location of the factory	1	1	1	1	0	1	1	1	1	1	**9**
	Dependency	**4**	**9**	**10**	**6**	**3**	**7**	**2**	**6**	**8**	**8**	

3.7 Level Partitioning the Final Reachability Matrix

Further with the help of reachability matrix, level partitioning is performed, and six different iteration tables were prepared. The hierarchy of parameters was decided using these iteration tables and further a green manufacturing framework was

Table 6 Reachability matrix

	Parameters/Element	Reachability Set	Antecedent Set	Intersection Set	Level
1	Type of Energy used	**1,2,5,8,9,10**	**1,2,3,4,5,6,7,8,10**	**1,5,8,10**	
2	Raw Material Manufacturing	**1,2,5,7,8,9**	**1,2,4,7,8,9,10**	**2,7,8,9**	
3	Apparel Manufacturing Technology	**1,3,5,8,9**	**3,4,6,8,9,10**	**3,8,9**	
4	Government Norms for the Industry or Regulatory Framework	**1,2,3,4,5,6,7,8,10**	**4,10**	**4,10**	
5	Procedure for Waste Treatment	**1,5,7,8,9,10**	**1,2,3,4,5,6,10**	**1,5,10**	
6	Economic Constraints	**1,2,3,5,6,7,8,9**	**4,6,9**	**6,9**	
7	Green Logistics (Packaging and Transportation)	**1,2,7,8,9**	**2,4,5,6,7,10**	**2,7**	
8	Competitive Strategies	**1,2,3,8,9**	**1,2,3,4,5,6,7,8,9,10**	**1,2,3,8,9**	**I**
9	Brand Building	**2,3,6,8,9**	**1,2,3,5,6,7,8,9,10**	**2,3,6,8,9**	**I**
10	Location of the Factory	**1,2,3,4,5,7,8,9,10**	**1,4,5,10**	**1,4,5,10**	

prepared. Table 6 presents the first iteration table for level partitioning, representing reachability, antecedent and intersection for each element.

3.8 *Hierarchy of Parameter*

Based on the outcome of six reachability iteration tables, parameters were divided into different layers. As mentioned in Table 6 (first reachability iteration table) parameters 8 and 9 are having common reachability, and the intersection set makes this group most "driven parameter", which means that no separate policies are required for these parameters as this might improve automatically with the improvements of other parameters. Similarly, parameter no. 4 is coming as most driver parameter and it requires a lot of policy interventions for improvement and have maximum impact toward the improvement of green practices of any manufacturing organization.

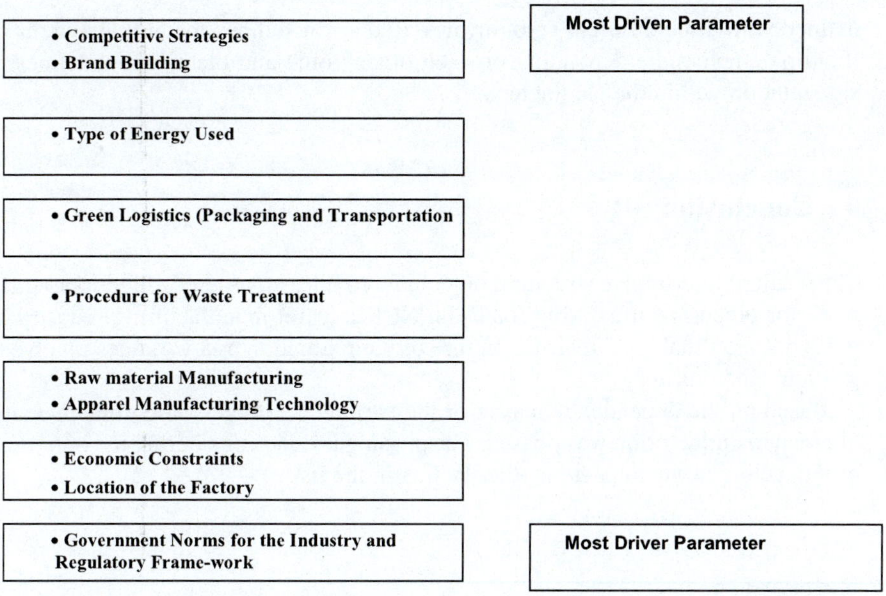

The probable phase-wise implementation of the framework is presented in Figure 3, which explains the relation between each parameter. The relationship is

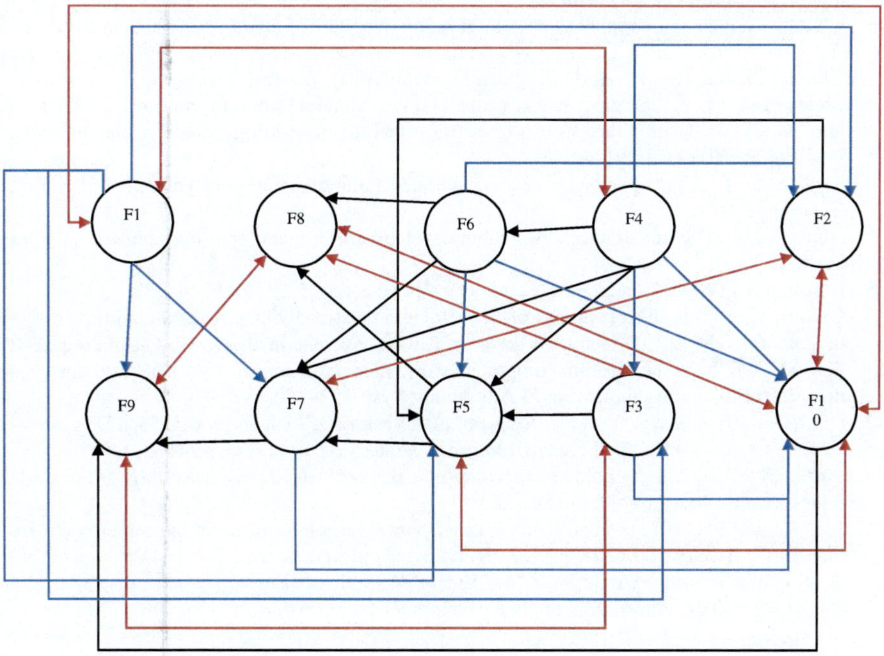

Fig. 3 System dynamics and proposed framework

defined with three different color arrows. Red arrow defines the both way relations (both parameters are depending on each other); blue and black arrow defines one side relation within the parameters.

4 Conclusion

The result of interpretive structural modeling provides an ordered, directional framework for green manufacturing for Delhi/NCR apparel manufacturing industry and gives decision makers a realistic picture of their situation and variables involved in green manufacturing.

Based on the dependency matrix of the various parameters, the implementation of green manufacturing was phased. The phasing is done considering the importance of the factor and its impact on other factors in the list.

References

1. Douglas J (2006) Building adaptation. J Clean Prod. Butterworth-Heinemann 99–110
2. International Panel of climate control (IPCC) (2009) Expert meeting on the sciences of alternative Metrics, Oslow, Norvey
3. Climate Change and Resource Sustainability- an overview of actuaries, Canadian Institute of Actuaries, 2015
4. Climate Change Report, Worls Research Institute (WRI), Zeneva, 2015
5. Mehta M (2014) V, Emerging issues in apparel trade. Apparel Export Promotion Council, Delhi
6. Das, M (2013) Performance measurement of green manufacturing criteria of Indian SME,s. Int J Eng Res Technol:2913–2920
7. Venkatesh VG (2015) Factor Influencing sucessfull implementation of green manfacturing. J Clean Prod:1–16
8. Adner R (2006) Evaluation of sustainable development in manufacturing industries. J Clean Prod
9. Bordoloi SA (2008) Design for control: a new perspective. Int J Prod:346–358
10. Gerrard JA (2007) Is European end-of-life vehicle legislation living up to expectation? Assesing the impact of the ELV Directive on green innivation and vehicle recovery. J Clean Prod:17–27
11. Tan XC (2008) A decision-making framework model of cutting tool selection for green manufacturing and its application. J Adv Manuf Syst: 257–260
12. Deif AM (2011) A system model for green manufacturing. J Clean Prod:1553–1559
13. Hoffman A (2000) Competitive environmental strategy. Island Press, pp 1564–1578
14. Shirley R (2014) A household carbon footprint calculator for islands: case study of the United States Virgin Islands Ecol Econ 80:8–14
15. Juan Cagiao BG (2011) Calculation of the corporate carbon footprint of the cement industry
16. Shailee G, Acharya DJ (2014) A review on evaluating green manufacturing for sustainable development in foundry industries. Int J Emerg Technol Adv Eng
17. Kannan G (2010) Analyzing supplier development criteria for an automobile industry. Ind Manag Data Syst:43–62
18. Eryuruk SH (2012) Greening of the textile and clothing industry. FIBRES & TEXTILES in Eastern Europe

19. Guo-Ciang Wu J-HD-S (2014) The effects of GSCM drivers and institutional pressures on GSCM practices in Taiwan's textile and apparel industry. Int J Prod Econ:618–636
20. Baskaran V (2014) Indian textile suppliers' sustainability evaluation using the grey approach. Int J Prod Econ:647–658
21. Caniato F (2015) Environmental sustainability in fashion supply chains: An exploratory case based research. Int J Prod Econ:659–670
22. Ministry of Textiles (2016) Annual report-2016. Ministry of Textiles, New Delhi
23. Ramachandra T V, KS (2015) Assessment of Carbon Footprint in Different Industrial Sectors. Springer
24. Sharma DM (2016) Comprehensive study on green house gases (GHGs) in. IIT Kanpur, Kanpur
25. Devaraja D (2012) Indian textile and garment industry. Bangalore. Indian Council of Social Science- Ministry of HRD, Govt. of India
26. Chattopadhyay S (2015) Efficiency of Indian garment manufacturing units in the post MFA period. J Ind Stat:58–75

An Efficient Supply Chain in Fast Fashion Through IoT

Komal Gahletia

1 Introduction

One of the most profitable markets in the world is the fashion and lifestyle industry, and it is defined to be a billion-dollar industry employing millions of professions all around the world [1]. Fashion industry is one of the most dynamic supply chains in the modern world and due to this nature, there are new challenges and many opportunities presented. With the fashion industry going global, consumers all over the world have been affected by this fast and constantly evolving industry disrupting consumer behavior in the global market.

We need to address the problem of distributing over time a limited amount of inventory across all the stores in a fast-fashion retail network [1]. The challenges specific to that environment include very short product life-cycles, and store policies whereby a reference is removed from display whenever one of its key sizes stocks out. The fast-fashion retail model gives rise to several important and novel operational challenges such as the problem of distributing over time a limited amount of merchandise inventory in a retail paradigm (short product life-cycles, store inventory display policies) do justify new approaches. Additionally, the retail operators are increasingly working in multi-channel and omni-channel environment, wherein the retail chains have to make their products available over different distribution channels like own websites, mobile applications (m-apps), third-party websites (e.g. Jabong, Myntra, Nykaa etc.) and also showrooms.

A sales channel serves two primary functions: delivering information and products to customers. Omnichannel retailing allows for the decoupling of these two functions because consumers can learn about products through channels that differ from those used to purchase them. This decoupling requires a far more sophisticated inventory and supply chain operation, as well as integration of all customer touchpoints, to

K. Gahletia (✉)
School of Management, World University of Design, Sonepat 131029, Haryana, India
e-mail: Komal.gahletia@wud.ac.in

A. Majumdar et al. (eds.), *Functional Textiles and Clothing 2020*,
https://doi.org/10.1007/978-981-15-9376-5_15

match fast-moving supply and demand [2, 3]. This decoupling has occurred because, in effect, omnichannel retailing blurs boundaries between the channels—in stark contrast to the more conventional multichannel approach followed by most retailers over the last decade.

So far, the omnichannel strategy for most retailers has hinged on developing a presence on social networks (Facebook, Instagram, etc.) and then devising ways for customers to shop on their smartphones. However, retailers are now beginning to deploy IoT devices and a new generation of software tools (World Economic Forum 2017).

In this competitive business world, most of the supply chains are struggling to sustain competitively in the global supply chain due to its increasing complexity in each phase of the supply chain operations. Companies have to be smarter by incorporating the necessary technologies to be more competitive and sustain in the global supply chain such that the processes can be better managed and automated where necessary [4]. Most companies fail due to poor integrations of technology in their supply chain. It is vital for companies to adopt to the changing nature of the digital supply chains and embrace Industry 4.0, leveraging internet of things (IoT)—a term coined by Kevin Ashton in 1999 (Fig. 1).

All industries, including fashion, are expecting that the IoT will make real quantifiable impact [5] that can be quickly translated into positive ROI for the business, and equally a positive return for consumers. The internet of things provides solutions based on the integration of various information technologies, which include hardware and software used to store, retrieve and process information, and communications technology, which includes electronic systems used for communication between individuals or groups of devices [6].

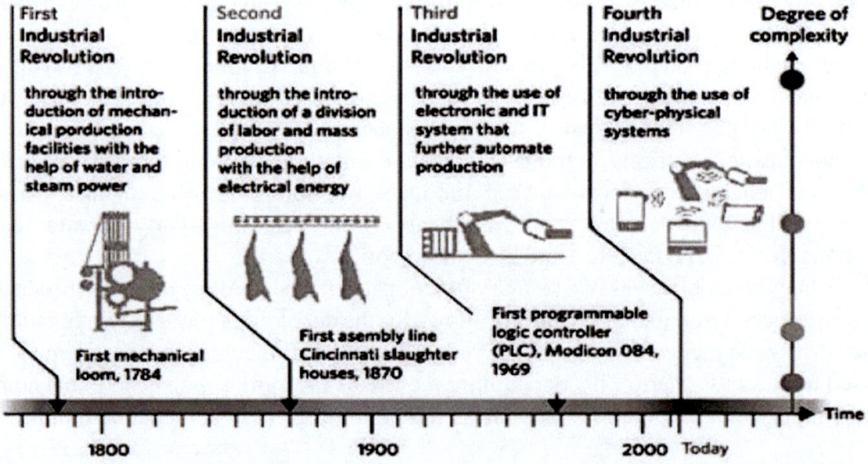

Fig. 1 The stages of industrial revolution in human history. *Source* "TheConvergenceofInformationand CommunicationTechnologiesGains Momentum," [12]. https://www3.weforum.org/docs/GITR/2012/G ITR_Chapter1.2_2012.pdf, accessed 25–09–2016

Objects and devices are more intelligent and they make themselves recognizable and they obtain intelligence by making various decisions as per their configured process and this is mainly due to the fact that they can communicate between them with a common understanding protocol. These objects and devices can access information that has been aggregated by physical objects, devices and sensors, or they can be components of an intricate network of services. This transformation is enabled with the emergence of cloud computing capabilities and the transition of the internet toward IPv6 protocol with an almost unlimited addressing capacity which was lacking in IPv4.

The internet of things provides solutions based on the integration of various information technologies, which include hardware and software used to store, retrieve and process information, and communications technology, which includes electronic systems used for communication between individuals or groups of devices.

The fashion industry has been constantly evolving with respect to the distribution channels, reducing the time from conceptualizing to manufacture to ultimate sales, their communication with their customers and business models. Traditionally, fashion apparel retailers used their capability of forecasting consumer demand and fashion trends (known as ready-to-wear) long before the actual time of consumption in order to compete in the market [7]. The fashion apparel industry developed an infrastructure of outsourcing manufacturing and processes to offshore places with low labor costs, thereby resulting in a substantial cost advantage [8]. However, recent years have seen fashion retailers compete with others by ensuring speed to market with their ability to provide rapidly the fashion trends revealed by fashion shows and runways. According to Taplin (1999) [9], such retailers could be credited with the adoption of 'quick fashion' that is an outcome of an unplanned process on the reduced time gap between designing and consumption on a seasonal basis. This 'quick fashion' is the precursor of 'fast fashion' model.

The key defining feature of 'fast fashion' retail model lies in novel product development processes and supply chain architectures relying more heavily on local cutting, dyeing and/or sewing, in contrast with the traditional outsourcing of these activities from developing countries. While such local production obviously increases labor costs, it also provides greater supply flexibility and market responsiveness. Indeed, fast-fashion retailers offer in each season a larger number of references produced in smaller series. The benefits of such a move is that products offered by fast-fashion retailers during the selling season may result from design changes decided after the season has started as a response to actual sales information, which considerably eases the matching of supply with demand.

Today's fashion market is highly competitive and the constant need to 'refresh' product ranges means that there is an inevitable move by many retailers to extend the number of 'seasons', that is, the frequency with which the entire merchandise within a store is changed. With the emergence of small collections of merchandise, fashion retailers are encouraging consumers to visit their stores more frequently with the idea of 'Here Today, Gone Tomorrow'. This indicates a shorter life-cycle and higher profit margins from the sale of fast selling merchandise, skipping the markdown process altogether. In addition, the desire to have variety and instant gratification is motivating consumers to prefer retailers espousing fast-fashion model.

The fast-fashion retail model gives rise to several important and novel operational challenges. The democratization of the fashion shows, wherein the conscious consumers are exposed to exclusive designs and styles inspired from fashion shows and internet, and expect and demand the same in shortest time possible is one such challenge. The retailers are expected to reduce their lead times and provide the designs displayed on the ramp to the distribution centers within 3–5 weeks. This led to the fashion apparel industry shift from forecasting future trends to using real-time data to understand the needs and desires of the consumers [10].

2 Research Problem

In fast-fashion online retail, it is imperative that the stocks information is available throughout the supply chain. Unlike in brick and mortar showroom, in virtual retail there is no human intervention. Here the systems interact and disperse the information across the retail channel. Therefore, the company adhering to fast-fashion philosophy needs the fastest and efficient way to communicate the information to relevant nodes/stakeholders.

The generic problem of allocating inventory from a central warehouse to several locations satisfying separate demand streams has received much attention in the literature. But the usage of IoT-enabled supply chain in an omnichannel world is missing, wherein different channels have different requirements.

3 Aim of the Study

The primary goal of this study is to improve inbound and outbound operations to better manage and optimize and automate operations in an ERP system through the use of IoT technology. To achieve the purpose, we apply IOT in SCM for making connection between supply chain entities and processes, identifying products and goods automatically, tracking flow of products at each stage, providing a complete information during the entire life-cycle of products, and achieving transparency of supply chain system to overcome challenges of traditional SC.

4 Methods

The methodology consists of literature review, primary interviews with the companies engaged in fast-fashion retail and online retail. The authors have identified logistics companies who are engaged in the supply chain operations of apparels for different

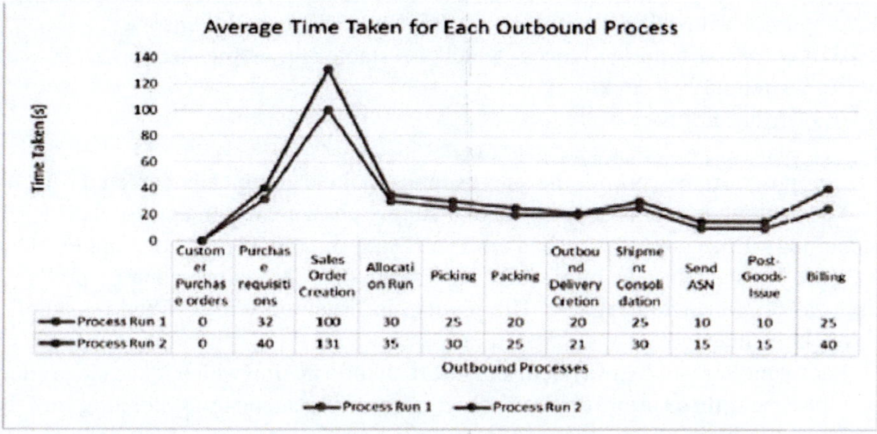

Fig. 2 The graphical representation of average time taken for outbound process

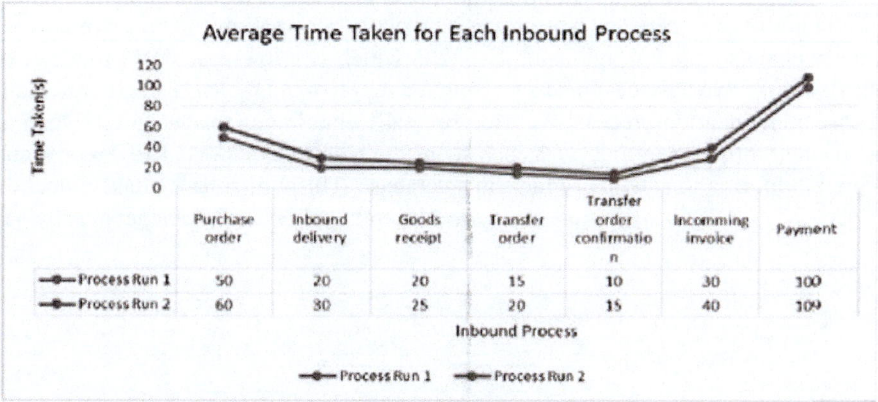

Fig. 3 The graphical representation of average time taken for each inbound process

companies. They were extensively interviewed and insights were taken from them. Based on those insights the following charts were developed (Figs. 2 and 3).

5 Results

- Enable process of supply chain [11].

The enabling technologies of internet of things usually consists of four major layers, which are as follows:

1. Layer for data collection, which use RFID technology and sensors,
2. Layer for transmission process which use stable and mobile networks,
3. Layer for service, and.
4. Layer for interface.

The proposed framework for the smart supply chain is graphically depicted (Fig. 4):

For tracking products at each stage of supply chain management, we used RFID technology. Each product attached with RFID tag and the RFID reader and Esp8266 are used for scanning the products at each stage of SCM. After scanning products, the tag ID uploads to database. The product information will be filled by supplier through the manager.

Each supplier will login system and insert all information which relates to product or service and these information saves in system. The manager by logging into the system can obtain all the required information about supplier and his/her product. He/she will send the product status and final decision to system. Each product attached with Tag ID, the RFID reader scans the product and send Tag ID to database. Through using RFID technology all information about products will be available, such as production, expire date and warranty period. This information can be shared across supply chain stages through using Esp8266 which is a low-cost Wi-Fi module. In the first stage, we have designed a website for supplier and manager in order to facilitate communication process. We used several technologies related to IOT such as RFID tag for tracking products at each stage and scanning them via RFID reader and store all information about products in a database. This will enhance data collection process. These information are shared between suppliers and managers easily via

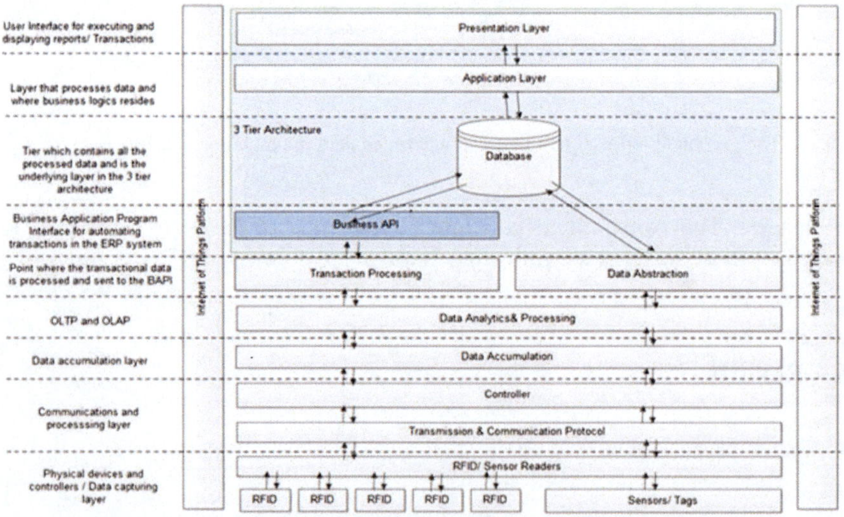

Fig. 4 The proposed framework depicting the various layers of smart supply chain. "The proposed framework for the smart supply chain is graphically depicted."

using Esp8266 which is a low-cost Wi-Fi module. This will achieve system transparency. Both manager and supplier can obtain product information from system database. In the second stage, according to the obtained information from database, the manager will evaluate supplier's product and select only the high-quality products. The selected products are purchased from the supplier and then stored in the warehouses. The supplier in this phase can access to system via entering username, password and track product status (accept, reject). In the final stage, after evaluating supplier's products and selecting the best, then a purchase process should be executed. After purchasing and processing products, a smart transportation system should be available for distribution process. We used global positioning system (GPS), geographic information system (GIS), and sensors technology to track vehicle location and ensure safety of on-board products. According to order information, the products will be delivered to customers. The obtained information from customer order is established via smartphones technology.

6 Conclusions

Fast fashion is a concept that will continue to affect the fashion apparel industry over the next decade and will have a direct effect on the way consumers purchase and react to trends.

The proposed system automates the identification process of products, trace and track products globally, achieves transparency, reduces time and cost, and then will achieve customer satisfaction. The designed website between company managers also enhances coordination process, makes suppliers able to find selling information of their products easily through entering system with their username and password. Our website removes middlemen via direct communication between suppliers and managers via system, and then it increases profit for both manager and supplier.

Not all retailers are equally predisposed to implementing IoT devices. Retailers that sell their own brand can easily set up RFID tagging, but those that stock private labels and/or sell items from multiple brands face greater challenges. That is, using RFID in such cases requires either a mandate issued to suppliers or the tagging of items at the retailer's distribution center.

References

1. Bhardwaj V, Fairhurst A (2010) Fast fashion: response to changes in the fashion industry. Int Rev Retail Distrib Consum Res 20:165–167
2. Bell DR, Gallino S, Moreno A (2014) How to win in an omnichannel world. MIT Sloan Manag Rev 56(1):45
3. Verhoef P, Kannan PK, Inman J (2015) From multi-channel retailing to omni-channel retailing. J Retail 91(2):174–181

4. Edwards P, Peters M, Sharman G (2001) The effectiveness of information systems in supporting the extended supply chain. J Bus Logist 22(1):122
5. Industrial Internet of Things (IIoT) and Its Impact on the Design of Automation Systems. https://www.maximintegrated.com/en/appnotes/index.mvp/id/6142. Accessed 27 Sept 2016
6. Fernie J, Sparks L (eds) (2009) Logistics and retail management: emerging issues and new challenges in the retail supply chain, 3rd edn. Kogan Page, London; Philadelphia
7. Guercini S (2001) Relation between branding and growth of the firm in new quick fashion formulas: analysis of an Italian case. J Fashion Mark Manag 5(1):69–79
8. Barnes L, Lea-Greenwood G, Tyler D, Heeley J, Bhamra T (2006) Supply chain influences on new product development in fashion clothing. J Fashion Mark Manag: An International Journal
9. Taplin IM (1999) Continuity and change in the US apparel industry: a statistical profile. J Fashion Mark Manag 3:360–368
10. Jackson T (2012) The process of fashion trend development leading to a season. In: Fashion marketing: contemporary issues. Routledge, London, pp 142–155
11. Borgia E (2014) The Internet of Things vision: key features, applications and open issues. Comput Commun 54:1–31
12. World Economic Forum (2017) Shaping the future of retail for consumer industries. www3.weforum.org/docs/IP/2016/CO/WEF_AM17_FutureofRetailInsightReport.pdf